Carbon-Energy Taxation

Carbon-Energy Taxation

Lessons from Europe

Edited by

Mikael Skou Andersen and Paul Ekins

OXFORD
UNIVERSITY PRESS

Great Clarendon Street, Oxford OX2 6DP

Oxford University Press is a department of the University of Oxford.
It furthers the University's objective of excellence in research, scholarship,
and education by publishing worldwide in

Oxford New York

Auckland Cape Town Dar es Salaam Hong Kong Karachi
Kuala Lumpur Madrid Melbourne Mexico City Nairobi
New Delhi Shanghai Taipei Toronto

With offices in

Argentina Austria Brazil Chile Czech Republic France Greece
Guatemala Hungary Italy Japan Poland Portugal Singapore
South Korea Switzerland Thailand Turkey Ukraine Vietnam

Oxford is a registered trade mark of Oxford University Press
in the UK and in certain other countries

Published in the United States
by Oxford University Press Inc., New York

British Library Cataloguing in Publication Data

Data available

Library of Congress Cataloging in Publication Data

Data available

Typeset by SPI Publisher Services, Pondicherry, India
Printed in Great Britain
on acid-free paper by
the MPG Books Group, Bodmin and King's Lynn

ISBN 978-0-19-957068-3

1 3 5 7 9 10 8 6 4 2

Preface

Taxes on carbon and energy remain an important instrument to curb greenhouse gas emissions, in particular for those emitters and installations that are not covered by emissions trading for CO_2. Even emitters under the European Union's emissions trading system are likely to continue to face taxes on their use of energy for reasons indicated in the EU's Energy Taxation Directive and relating to the functioning of the internal market as well as to concerns for security of energy supply.

Carbon-energy taxes can be applied both at the level of individual EU member states and collectively by the EU as a whole, while revenues can be recycled to reduce taxes on labour. In this book we explore the impact such taxes, introduced under unilateral environmental tax reforms, may have on competitiveness, as well as on efficient use of energy and on reductions in related carbon emissions.

The research underlying this book was made possible with a grant to the project COMETR (The Competitiveness Effects of Environmental Tax Reform) from the European Union's Sixth Framework Programme for Research under the Scientific Support to Policies programme.

While the initial motive for research on the topic of environmental taxation was the EU's Lisbon strategy for improving Europe's competitiveness *vis-à-vis* other major trading blocs, the COMETR research proved timely in relation to the revived climate strategy which the European Council decided to embark upon in March 2007 in response to the Fourth Assessment report from IPCC, the Intergovernmental Panel on Climate Change.

Only ten days after Europe's Heads of State had announced a new and more vigorous climate strategy at the Summit on 9 March 2007, committing EU to a 20–30 per cent reduction in greenhouse gas emissions by 2020, more than 500 tax administrators from member states gathered in Brussels for the annual Tax Forum hosted by the Commissioner for Taxation and Customs Union, Laszló Kovács, to consider the topic of sustainable taxation. At this event, participants were presented with key findings from the research on carbon-energy taxation published in this book. Subsequently, results from COMETR helped inform preparation of the European Commission's climate policy package of January 2008, in particular with respect to the implications of pre-existing carbon-energy taxation.

The final climate policy adopted by the European Council and Parliament in December 2008, in particular the delay in the introduction of full auctioning of permits in the EU Emissions Trading Scheme, shows the continuing resonance of arguments about competitiveness impacts from energy and climate policy. We hope that the material in this book will help inform policy-makers in their further consideration of such arguments.

In light of the findings presented, it seems likely that exemptions from the auction of carbon allowances and from energy taxation will be subject to more rigorous scrutiny than before, with regard to the actual implications for competitiveness. The revised environmental guidelines for approval of state aid, issued by the European Commission in April 2008, play a key role in this regard.

There are a number of good reasons why carbon-energy taxation could in the future be combined with trading of carbon emissions allowances, as explained in Chapter 9, 'Carbon Taxes and Emissions Trading: Issues and Interactions' by Paul Ekins. Taxes can help place a floor under the volatile carbon pricing implied by trading schemes, which is important for giving an assurance to low-carbon investors of a minimum carbon price. Carbon-energy taxes can also play an important role in developing countries, such as China, that are unlikely to accept caps on their emissions, but which can reap the dividends of tax reform, for instance beginning with a tax related to the carbon-energy content of their export products.

The members of the research group behind COMETR came together for the purpose of the project and comprised six institutes from five different member states. Coordinated by National Environmental Research Institute (NERI), Aarhus University in Denmark, they included Cambridge Econometrics, the Economic and Social Research Institute in Dublin, the University of Economics in Prague, the Policy Studies Institute in London, and the Vienna Institute for International Economic Studies. The institutes had not previously joined forces, but some individual scholars had participated in a large concerted action on market-based instruments under the Fourth Framework Programme of Research, while there had been bilateral cooperation between Danish and Czech scholars in a pan-European framework on capacity building. We are grateful in particular to Professor Frank Convery, University College Dublin, for initiating some of the earlier exchange and networking that eventually led to our collaboration in the COMETR project.

In relation to the COMETR project, we are indebted to a number of partners and colleagues for input to and feedback on our research. We are especially grateful to academic colleagues who gave important responses to our draft reports, in particular Professor Michael Landesmann, University of Linz, Professor Bernd Meyer, University of Osnabrück, and Professor Georg Müller Fürstenberger, University of Trier. Senior national experts from tax administrations provided important detail to help our understanding of exemptions and the implications in terms of actual tax burdens for industries, and we would like to thank in particular Susanne Åkerfeldt (Sweden), Petri Malinen (Finland), and Hans Larsen (Denmark). Last but not least, we are very grateful to Katri Kosonen in DG TAXUD and Ian Perry in DG Research, both of the European Commission, for their interest and good understanding of the nature of the research process. We also extend our thanks to Adela Tesarova in DG TAXUD for her helpful cooperation in relation to the Tax Forum, to Pierre Valette in DG Research for general project facilitation, to Hans Vos at the European Environment Agency for sharing his unique eco-tax expertise, and to Mark Hayden DG ECFIN for sensitizing us to the new context for the findings.

In preparing the manuscripts for publication we had excellent assistance from Carey E. Smith and Ann-Katrine Holme Christoffersen, NERI, Aarhus University.

Mikael Skou Andersen and Paul Ekins

Aarhus/London, December 2008

Contents

Contents

Contents

List of Figures

List of Tables

Abbreviations

BOF	basic oxygen furnace
CCA	Climate Change Agreement
CCICED	China Council for International Cooperation on Environment and Development
CCL	climate change levy
CCS	carbon capture and storage
CE	Cambridge Econometrics
CEEC	Central and Eastern European countries
CGE	computable general equilibrium model
CHP	combined heat and power
COMETR	competitiveness effects of environmental tax reforms
CO_2	carbon dioxide
CO_2-eq	carbon dioxide equivalents
CPI	consumer price index
CT	COMETR tax
CR	COMETR revenue recycling
DKK	Danish kroner
DM	Deutschmark
EEA	European Environment Agency
EFR	environmental fiscal reform
EMS	European Monetary System
EPA	Environmental Protection Agency
ESRI	Economic and Social Research Institute (Dublin)
ETR	environmental tax reform
ETS	emissions trading system
EU	European Union

EUR	Euro
E3ME	energy-environment-economy model of Europe
FIM	Finnish mark
FFL	fossil fuel levy
GDP	gross domestic product
GHG	greenhouse gases
GMM	generalized method of moments
GOS	gross operating surplus
HFL	Dutch guilders
HG	homogeneous goods
ICT	Information and Communications Technology
IEA	International Energy Agency
IO	input-output
IPCC	Intergovernmental Panel on Climate Change
IPPC	integrated pollution prevention and control
IRTS	increasing returns to scale
LPG	liquefied petroleum gas
MCA	marginal cost of abatement
NACE	nomenclature des activités économiques dans la Communauté Européene
NERI	National Environmental Research Institute
nes	not elsewhere specified
NFFO	non-fossil fuel obligation
NIC	national insurance contribution
NOx	nitrogen oxides
OECD	Organisation for Economic Cooperation and Development
OTAF	Office of Technology Assessments and Forecasts
PETRE	Productivity and Environmental Tax Reform in Europe
pp	percentage point
ppm	parts per million
PPP	purchasing power parity
RATS	regression analysis of time series
ROC	renewable obligation certificate
R&D	research and development
SEK	Swedish kroner

Abbreviations

SIT	Slovene Tolar
SO_2	sulphur dioxide
SSC	social security contributions
STAN	structural analysis database (OECD)
UK	United Kingdom
UNFCCC	United Nations' Framework Convention on Climate Change
TAR	Third Assessment Report (IPCC)
VAT	value added tax
WTO	World Trade Organization

Notes on Contributors

Mikael Skou Andersen is Research Professor of Environmental Policy Analysis at Denmark's National Environmental Research Institute (NERI), Aarhus University. His Ph.D. and Master's degrees were obtained from the Faculty of Social Sciences at Aarhus University, where he was previously Associate Professor in the Department of Political Science (1995–2000). He was a member of the Danish Minister of Taxation's Working Group on CO_2 (2005–6). He is involved in several large research projects on energy systems and air pollution, including AIRPOLIFE and CEEH, Centre for Energy, Environment and Health. His academic work focuses on comparative research on environmental and climate policies, accounting for the properties and effects of various policy instruments, including environmental taxes, as well as research to develop interdisciplinary collaboration between the natural sciences and social sciences, in particular to improve understanding of the external costs of pollution. He has published several books including *Market-Based Instruments for Environmental Management* (Edward Elgar, 2000, co-edited with R. Sprenger), and the monograph *Governance by Green Taxes* (Manchester University Press, 1994). He is currently a member of the Task Force on Economic Instruments and Energy Efficiency under the China Council for International Cooperation on Environment and Development.

Terry Barker MA (Edinburgh), MA (Cambridge), Ph.D. (Cambridge); Chairman and Consultant, Cambridge Econometrics. Dr Barker is the Chairman of Cambridge Econometrics, having founded the company in 1985. Since 2005 he has also been the Director of the Cambridge Centre for Climate Change Mitigation Research (4CMR), Department of Land Economy, University of Cambridge. In addition, he is a member of the editorial board of Economic Systems Research.

He was a co-ordinating lead author (CLA) for the Intergovernmental Panel on Climate Control (IPCC)'s Fourth Assessment Report, 2007, for the chapter on cross-sectoral mitigation. Previously he was CLA in the

Third Assessment Report, 2001, taking responsibility for the chapter on the effects of greenhouse gas mitigation policies on global energy industries. He was a member of the core writing team for the Synthesis Report Climate Change 2001.

From 2000 he instigated and worked on projects building a global E3 model (E3MG) with initial emphasis on modelling the E3 structures of China and Japan. Since 2004 he has been working as a member of a UK Tyndall Centre project to develop E3MG as a 20-region world model, designed to analyse GHG mitigation policies under endogenous technological change. In the 1990s he was appointed the project co-ordinator of the pan-European project developing and applying the E3 model for Europe (E3ME), partly funded by the European Commission, analysing energy and fiscal policies, including the equity effects of environmental fiscal reform.

His publications include *International Competitiveness and Environmental Policies* (with Jonathan Köhler; Edward Elgar, 1998) and 'The effects on competitiveness of co-ordinated versus unilateral fiscal policies reducing GHG emissions in the EU: an assessment of a 10% reduction by 2010 using the E3ME model', *Energy Policy*, 26/14, 1998, pp. 1083–98.

Paul Ekins has a Ph.D. in economics from Birkbeck College, and a BSc in electrical engineering from Imperial College (both University of London). He is Professor of Energy and Environment Policy at King's College London and from August 2009 will be Professor of Energy and Environment Policy at the Energy Institute at University College London. He was a Member of the Royal Commission on Environmental Pollution from 2002–8. He is the Director of the UK Green Fiscal Commission, which is exploring the prospects for and implications of large-scale environmental tax reform in the UK. His academic work focuses on the conditions and policies for achieving an environmentally sustainable economy, with a special focus on energy policy, and the modelling of the energy system, on innovation, on the role of economic instruments such as environmental taxes, on sustainability assessment, and on environment and trade. His books include *Economic Growth and Environmental Sustainability: the Prospects for Green Growth* (London: Routledge, 2000). Recently co-edited books include *Understanding the Costs of Environmental Regulation in Europe* (Cheltenham: Edward Elgar, 2009), *Trade, Globalization, and Sustainability Impact Assessment: A Critical Look at Methods and Outcomes* (London: Earthscan, 2009), and *Hydrogen Energy: Economic and Social Challenges* (London: Earthscan, 2009). In 1994 Paul Ekins received a Global 500 Award 'for outstanding environmental achievement' from the United Nations Environment Programme.

Martin K. Enevoldsen holds an MA and a Ph.D. in political science from Aarhus University (Denmark). Martin Enevoldsen was until 2005 associate professor in the Department of Political Science, Aarhus University and is the author of *The Theory of Environmental Agreements and Taxes: CO₂ Policy Performance in Comparative Perspective* (Edward Elgar, 2005). Martin Enevoldsen is currently manager in the strategy department of Deloitte Business Consulting in Copenhagen.

John Fitz Gerald is a Research Professor with the Economic and Social Research Institute (ESRI) in Dublin. As a macroeconomist, he is responsible for the ESRI's Medium-Term Review. He is Director of the ESRI's Energy Policy Research Centre and he has published widely on the economics of global warming and on regulatory reform in the energy sector. He is a member of the National Economic and Social Council and of the EU Group of Economic Analysis, which provides advice to the President of the EU Commission on matters of economic policy.

Jirina Jilkova is a Professor of Economics in the Department of Environmental Economics and Executive Director of the Institute for Economic and Environmental Policy, at the University of Economics, Prague. She is Advisor to the ministry (advising on environmental policy), in the Ministry of Environment of the Czech Republic. She has conducted numerous research projects, both national and international, dealing mainly with environmental economics and policy, rural development, and agricultural economics and policy. Such projects include: Domestic emissions trading system in the Czech Republic: options for an implementation framework, OECD, 2001, Methodology of economic impacts assessment of the environmental legislation, Ministry of Environment of the Czech Republic, 1998–2001, and Assessment of externalities of agricultural production, Ministry of Environment of the Czech Republic, 1999–2001. She has published many papers, articles, and books.

Sudhir Junankar is an Associate Director responsible for Cambridge Econometrics' (CE's) UK Energy-Environment Service. Sudhir Junankar manages the use of MDM, CE's energy-environment-economy model of the UK economy.

Recent projects he has managed at CE include: a project for the European Commission on the impacts of possible changes to the EU Emissions Trading Scheme, and a project reviewing the EU Energy Taxation Directive for the Taxation Directorate of the European Commission. He is currently leading a project for the Anglo-German Foundation, investigating the major research issues related to resource productivity and environmental tax reform (ETR) and sustainable growth in Europe. All four projects are based on the use of CE's E3ME model. Before joining CE, Sudhir Junankar was Associate Director, Economic Analysis at the Confederation of British Industry, where he worked for 14 years. He had primary responsibility for the team of economists and survey experts which undertake the CBI's highly regarded suite of business surveys of the private sector, and for distilling economic intelligence from these surveys to brief business and government leaders. He was responsible for the CBI's quarterly short-term forecasts of the UK economy, based on its econometric model. He represented the CBI at various international organizations including the European Commission, the Forecasting Group of the European Employers' Confederation, and OECD advisory bodies.

Mary J. Keeney is a research economist at the Central Bank and Financial Services Authority of Ireland. She gained her Ph.D. from Trinity College Dublin (TCD). She worked on natural resource, energy, and environmental policy issues while at the Economic and Social Research Institute (ESRI), Dublin. Prior to this, she contributed to the ESRI Tax Benefit model (SWITCH) and worked on related topics of income distribution and poverty measurement. She has collaborated on a number of research projects with European institutions, and is a member of several European research networks. Present areas of research relate to issues of household finances and wealth, financial capability, and the inflationary effects of firm-level decision-making. She is also a part-time lecturer at the Department of Economics at TCD.

Alexandra Miltner was a research fellow in the environment group at the Policy Studies Institute from 2005–7. She is an environmental economist with a strong quantitative background. Prior to joining PSI, she worked as a research assistant at the Yale University Department of Economics and at the Law School studying environmental issues. She holds an MA degree in Economics from Queen's University (Canada).

Hector Pollitt specializes in the application of econometric techniques to large, disaggregated data sets to interpret historical experience, to simulate the impact of alternative policy options, and for forecasting.

He leads on the operation and development of Cambridge Econometrics' large-scale European econometric model, E3ME, and on the company's contribution to developing the global E3MG model. He also contributes the same expertise to the similar UK model, MDM-E3. These responsibilities include overseeing the maintenance and development of the models' extensive time-series and cross-section databases, with detailed sectoral and regional disaggregation, and the estimation and updating of the models' parameters. He leads on the application of E3ME in impact studies and forecasting exercises, notably for energy-environment-economy analysis and for detailed sectoral analysis.

Anders Ryelund MA Political Science (Aarhus University, Denmark). Anders Ryelund worked as a junior researcher at the National Environmental Research Institute, Aarhus University. His research activities included analysis of environmental taxes reforms. Anders Ryelund currently works as an administrative officer at the health planning department in Region Midtjylland, Denmark.

Roger Salmons is a freelance environmental economist and a Visiting Research Fellow in the Environment Group at PSI. He has been actively involved in research on environmental policy issues since 1996; working on collaborative research projects with institutions throughout Europe and participating in several European research networks. Prior to this, he spent 14 years working in the commercial sector. He has a Ph.D. in economics from University College London and his primary research interests relate to the interaction between environmental policy and economic performance, and the design and evaluation of market-based environmental policy instruments. In addition to his research activities, he has acted as a consultant to government departments and agencies in the UK, to the European Commission, and to the OECD.

Susan Scott heads the ESRI's Environment Policy Research Centre, where she has coordinated and worked on studies commissioned by Government, the EU, and private bodies. She managed the project for the Department of the Environment, which was subsequently published as the book, *The Fiscal System and the Polluter Pays Principle* (co-authored with A. Barrett and J. Lawlor, Ashgate, 1997). She has organized conferences and workshops featuring national and international speakers, including: Economic instruments supporting environmental policy (1996), Taxation of excess nutrient use in agriculture (1996), Environment and EU Treaty revisions: The IGC Review (Institute of European Affairs, 1997), Green and bear it? Implementing market-based policies for Ireland's environment (2001).

The ESRI has managed and coordinated a number of major European research networks, including CHANGEQAL and CATEWE.

Stefan Speck is an environmental economist and holds a Ph.D. in economics from Keele University. He worked for several years at the Regional Environmental Centre for Central and Eastern Europe in Hungary, where he gained a thorough understanding of recent developments in the implementation of economic instruments in Central and Eastern European countries. He was also responsible for a project analysing environmental funds in the region. In addition, he compiled a database on economic instruments in EU member states funded by the European Commission and published several articles in the field of environmental taxation and competitiveness. Dr Speck contributed to a study evaluating the effectiveness of environmentally related taxes and charges in European countries. His experience also includes work for projects funded by the Danish Environmental Protection Agency (DEPA), the EC, the OECD, and the CPB Netherlands Bureau for Economic Policy Analysis. Speck currently works for Kommunalkredit in Austria.

Philip Summerton is an applied economist whose main specialization is in the application of Cambridge Econometrics' (CE's) large-scale energy-environment-economy models to analyse E3 issues in the UK and the rest of Europe. He leads the modelling and data analysis for the maintenance and development of CE's large-scale econometric models, notably MDM-E3 (the multisectoral dynamic energy-environment-economy model for the UK), and he also contributes to E3ME (energy-environment-economy model for Europe) and E3MG (energy-environment-economy global model).

Part I

Pricing of Carbon in Europe

1

Carbon-Energy Taxation, Revenue Recycling, and Competitiveness

Mikael Skou Andersen[1]

1.1 Introduction

Conceptual interest in carbon-energy taxation emerged across Europe in recognition of serious environmental challenges, notably those high-lighted in the first reports on the risks associated with greenhouse gases and global warming. Upon their return from the ground-breaking Toronto climate conference in 1988, several governments pledged to reduce CO_2 emissions by up to 20 per cent and to this end countries such as Finland, Sweden, and Denmark were soon to introduce unilateral CO_2 taxes. The Netherlands and also Slovenia followed suit and at the end of the 1990s key EU member states, Germany and UK, had introduced similar tax measures for greenhouse gases too. In countries as diverse as New Zealand, Japan, and Italy, carbon-energy taxation appeared on the political agenda and governments began to grapple with understanding the economic implications.

Carbon-energy taxation was conceived primarily as a policy instrument to curb carbon emissions, but in order to tackle the associated economic challenges, the idea of undertaking more comprehensive environmental tax reforms, in which the tax burden is shifted away from labour and on to energy and pollution, began to emerge (Nutzinger and Zahrnt, 1990). While carbon-energy taxation applied in isolation faced opposition due to concerns about the potential negative economic impacts, the wider

[1] Mikael Skou Andersen, Professor, Department of Policy Analysis, National Environmental Research Institute, Aarhus University, Denmark.

3

principle of environmental tax reform (ETR) offered a new macroeconomic climate in which the shift in the tax burden allowed for more immediate benefits in addition to long-term containment of climate change (O'Riordan, 1997). By lowering taxes on labour in return for taxation of carbon as part of a more comprehensive and revenue-neutral tax reform, there would be opportunities not only to tackle negative economic impacts but also to improve employment, while setting out on a trajectory of 'greener' growth.

In practice the idea to shift taxation from 'goods' to 'bads'—from labour to pollution—has not proved as compelling as it is simple. Neither the European Commission nor the White House administration have managed to implement the carbon tax proposals they conceived in the early 1990s. Nevertheless, starting with Sweden's tax reform in 1989, and later fuelled by a 1993 White Paper from Commission President Jacques Delors, a number of European countries have, in the absence of the unanimity required for EU harmonization, incrementally altered their tax systems in the direction of ETR (Vehmas, 2005). Altogether revenues of more than 25 billion EUR have been shifted from tax bases of labour to carbon-energy. Meanwhile a 2003 Energy Taxation Directive now provides some minimum fuel and power tax rates for Europe, while in North America, Canadian provinces (Quebec and British Columbia) have begun to implement ETRs in a piecemeal fashion (Duff, 2009).

With regard to environmental implications, there is little controversy that attaching a price to emissions will have the expected impacts and empirical studies are now available which substantiate expectations (Bruvoll and Larsen, 2004; Rapanos and Polemis, 2005; Enevoldsen, 2005). However, in a world where trade barriers have been removed in pursuit of free trade, unilateral tax shifts inevitably raise concerns about competitiveness impacts (Smith and Sims, 1985). In view of some draconian proposals to use taxes to double or triple energy prices (von Weizsäcker and Jesinghaus, 1992), competitiveness concerns are intuitively strong and it is hardly surprising that the potential impacts on competitiveness remain the key issue in both the academic and policy debate on ETR (Barker and Köhler, 1998; Smith, 2003).

In reality, most ETRs have adopted cautious and incremental approaches to increasing carbon-energy taxes, while at the same time lowering social contributions or labour taxes to mitigate the competitiveness impacts. From the very beginning, the principle of *revenue neutrality* has been at the heart of deliberations, implying that the methods used for shifting taxes are key to the possible success of ETR. In many cases, concerns presented on competitiveness grounds refer to the budget-economic implications of ETR for individual firms, neglecting to consider the recycling of tax revenue not to mention the benefits attached to energy efficiency improvements—as pollution and resources each come with a price tag.

Nevertheless even an ETR with associated tax shift will produce some structural effects, with some companies winning and others losing. Rather than the short-term impact on individual companies, however, it is the overall impact on a country's competitiveness which should be in focus (Esty and Porter, 2001).

The European Commission in its annual competitiveness report defines competitiveness as 'a sustained rise in the standards of living of a nation and as low a level of involuntary unemployment as possible' (EC, 2004: 17). Less succinctly, the OECD defines competitiveness as 'the degree to which a country can, under free and fair market conditions, produce goods and services which meet the test of international markets, while simultaneously maintaining and expanding the real incomes of its people over the longer term' (OECD, 1993: 237).

Whether ETR may actually contribute to *improving* economic performance continues to be hotly contested. However, if we think of economic performance as sustaining and expanding 'real incomes' (cf. the EU and OECD definitions above), we are urged to focus our attention more comprehensively on the possible dynamic impacts and their implications in particular for competitiveness. It is the mission of the present volume to address this debate. In the following section, claims on ETR put forward in the economic literature are summarized in order to lay out the logic underlying our research and enquiries reported in subsequent chapters.

1.2 The Porter hypothesis on the relationship between environmental regulation and competitiveness

Harvard economist Michael Porter argued in *The Competitive Advantage of Nations* (1990) that contrary to conventional wisdom, environmental standard-setting may actually be able to encourage innovation and hence improve competitiveness, in particular when regulatory standards anticipate requirements that will spread internationally (Ashford *et al.*, 1985; Porter, 1998: 187). This 'Porter hypothesis', reflecting and extending a broader literature on regulation and innovation, was proposed within a broader theoretical framework on competitiveness, where Porter argued that clusters of industries facing a 'diamond' of advantageous national circumstances would respond to pressures from outside by seeking more sophisticated sources of competitive advantage and 'ruthlessly' pursuing further improvements (van der Linde, 1993). Of the four corners in the diamond, 'environmental regulations' was the one that would particularly affect demand conditions for industries. They would create a market for new and greener products, while simultaneously altering the framework for industry rivalry.

Competitiveness, according to Porter, depends on the capacity of a nation's industry to innovate and upgrade, and it is pressures and challenges, in particular from strong domestic rivals, that lead companies to gain advantage against the world's best competitors.

From his few remarks in the book on the role of environmental regulations in relation to competitiveness, two subsequent articles were developed. Most often, the joint article with van der Linde (1995) is quoted in the literature, but in a previous essay on 'America's green strategy' (1991) Porter in fact elaborates more on the type of environmental regulations required in order to produce beneficial impacts improving competitiveness. He warns that the majority of previous environmental regulations have actually violated the principles for a positive impact on competitiveness, by having emphasized the application of specific pre-defined technologies, often end-of-pipe, rather than leaving room for adaptation, flexibility, and innovation. Instead of conventional command-and-control policies, standards should be enforced by market incentives, which also help contain control costs. It is this emphasis on the use of market-based instruments in environmental regulation which provides the stepping stone to ETR.

Porter and van der Linde point to six purposes which a well-designed, market-based environmental regulation can serve:

- directing attention to resource inefficiencies;
- raising corporate awareness and information gathering;
- providing more certainty to green innovators;
- overcoming organizational inertia and fostering creative thinking;
- improved learning, so that short-term losses can be reverted to long-term gains;
- inducing change, as benefits in many cases are incomplete until innovation-based solutions have been developed.

The Porter hypothesis fundamentally claims that in the longer term there will be innovation offsets from environmental regulations which will outweigh the costs imposed. Such innovation offsets can be either process- or product-oriented—the latter being regarded as the most promising in terms of radical shifts which improve competitiveness.

There was vigorous controversy in the 1990s over Porter and van der Linde's claims that there were 'low-hanging 10-pound notes' that had not been picked up by businesses (Jaffe *et al.*, 1995). Palmer, Oates, and Portney (1995) took strong issue with the view held by Porter and associates, in their opinion based too extensively on case studies and anecdotal evidence rather than on theoretical rigour. While the critics did not deny the existence of innovation offsets, they found them to be several orders of magnitude lower than the imposed costs of environmental regulation. The critics prefer to subject regulations to conventional cost-benefit analysis, where innovation offsets constitute only a portion of the social benefits concerned, and more generally favour a social contingency approach rather than one related to competitiveness.

Yet in response to these neoclassical critics, many supporters of the Porter hypothesis pointed out that organizational slack in company performance is in fact the subject of a large body of research literature, and that in real company management the challenge of identifying and harvesting the low-hanging 10-pound notes remains (Goodstein, 2003).

1.3 The double dividend debate

David Pearce (1991) directed attention to possible *double dividend* features of pollution taxes in the debate on policies and measures that followed the first report from the International Panel on Climate Change. Since pollution taxes serve to correct market failures, by definition they do not share the distorting properties of many other taxes. A fiscally neutral package can be adopted by replacing distorting income taxes or corporate taxes by carbon taxes—by means of revenue recycling.

The Pearce argument acknowledges that environmental regulations normally bring a first dividend of pollution control benefits. However, because of the long time horizon associated with these benefits and due to the inter-temporal dimension of many environmental issues, the potential existence and magnitude of a second and more short-term benefit for current generations should be afforded more attention, which appears relevant in the context of climate change and CO_2 taxation. The second dividend here brings increased social welfare; the principal route of effects being rising employment—if labour costs fall and energy and environment costs increase under ETR.

The 'double dividend' hypothesis can be regarded as a less bold version of the Porter hypothesis—it claims that social welfare, rather than competitiveness *per se*, is improved when an ETR is applied. In the context of the European Commission White Paper from 1993 the double dividend argument was also linked to an improvement in competitiveness, as the paper advocated taxing 'bads' (pollution) rather than 'goods' (labour), so as to improve overall efficiency. However, the case for improved social welfare seems to be founded both on gains from improved environmental quality and on winning market shares for new types of products, spurred by ETR.

Many economists had difficulties with the 'free lunch' implied in the double dividend argument, as well as with the rhetoric on the win-win options of environmental policy applied by its adherents. Goulder (1995) hence proposed to differentiate between *weak* and *strong* versions of the double dividend argument. The strong version of the double dividend hypothesis is the claim that an environmental tax which replaces another tax will always provide for a positive economic impact, as environmental taxes are non-distorting since they help to internalize some external costs, whereas taxes on labour are distorting to the economy. The weak version, on the other hand, merely focuses on the revenue-recycling aspect and claims that once environmental taxes have been introduced, using revenues to reduce other distortionary taxes is preferable to a lump-sum return of revenues; an uncontroversial standpoint. Finally, an *intermediate* version of the double dividend argument seems, according to Goulder's interpretation, to imply that whether overall social welfare will be improved as a result of ETR depends on the specific properties of the distortionary tax which is being replaced with an environmental tax—in other words, it depends on context and circumstances (cf. McCoy, 1997).

The intermediate version of the double dividend claim has been further extended by Bovenberg and de Mooij (1994). From a public finance position, they point to the existence of a possible 'tax interaction effect' that may countervail the revenue-recycling effect of ETR. In essence, the tax interaction effect will exceed the revenue-recycling effect, except under special circumstances with highly distortionary taxes. The mechanism of the tax interaction effect is that the environmental taxation causes commodity prices to increase, lowering the real value of after-tax income. It is claimed that because of labour supply elasticities, the net effect of the ETR will usually be negative, as the relief on income taxation provided by ETR is too small to offset the price increases.

This finding hinges on the crucial assumption that income taxation a priori minimizes the excess tax burden (Weinbrenner, 1999). It also hinges on the assumption that ETR is introduced on top of existing environmental taxes or regulations that already internalize all externalities.

One important modification to the tax interaction effect occurs if the ETR involves a direct lowering of employers' social contributions, so that no or only marginal price changes will result (Parry, 1995). In fact the swap between environment taxes and social security contributions is one that has been practised in several ETRs.

However, there is much to suggest that many of the analyses which focus on the tax interaction effect are too stylized and restrictive. Bovenberg and de Mooij's first article was based on a static model. In a second article (1997), where they explore the relationships in the context of a dynamic model, the findings are relaxed somewhat: if the ETR leads to lower regulatory pressure on companies then a double dividend may arise.

Nielsen *et al.* (1995) explore the double dividend hypothesis with a dynamic model that includes unemployment. They show that unemployment will be reduced if a pollution tax is introduced. In this case, the tax interaction effect also influences the value of the unemployment benefit, causing more unemployed to enter the labour market. The overall effect on the rate of economic growth could, however, become negative.

Goodstein (2003) generally questions the basic assumption of the tax interaction effect that higher prices will reduce labour supply. Quoting earlier empirical literature based on micro-data, this relationship is found to be ambiguous. Higher prices may lead to an increase in labour supply if dual earner families are considered. Workers may increase labour supply partly because they overestimate the reduction in family income generated by the price increases (cf. Gustafson and Hadley, 1989, quoted in Goodstein, 2003).

The controversy over the second dividend has become prominent because the first dividend has appeared to be highly uncertain and in any case will materialize only within an extended time-frame. The social welfare effect from introducing a carbon-energy tax hence depends less on the expected long-run environmental benefit and more on the benefits of reducing a pre-existing, distortionary tax, but net of any distortions generated by the environmental tax and the related tax interaction effects. With such a focus, the implications of a tax shift for the long-run competitiveness of the economy requires improved reflection and analysis.

1.4 What kind of efficiency are we talking about?

The problem with much of the debate on the double dividend hypothesis is that it implicitly frames the issue as one of simple *allocative* efficiency. The proposal in the Delors White Paper (1993) to shift taxation from labour to pollution and natural resources was indeed conceived within a conceptual framework of improved allocative efficiency resulting from a change in input factors. However, a number of authors argue that what Porter and others seem to be addressing should probably be regarded rather as *incentive* efficiency (Pearce, 2001).

In a landmark article, Berkeley economist Leibenstein (1966) proposes to distinguish what he terms 'X-efficiency' from traditional allocative efficiency. While allocative efficiency addresses the optimal combination of productive resources, X-efficiency addresses the optimal use of the individual factor of production. Leibenstein discusses whether labour is always used optimally, citing extensive evidence for productivity improvements achieved in the use of labour. The scope for such improvements would normally be assumed away by neoclassical theory's assumption of optimality and rationality in the management of firms. Yet, on the issue of monopoly regulation, the welfare improvements from X-efficiency could be justified theoretically and empirically to be of a much larger scale than simple allocative efficiency gains.

Leibenstein provides a number of reasons why managers and employees would prefer not to produce at the outermost bound of optimality, for example, to avoid the required effort and pain of full efficiency. 'It is one thing to purchase or hire inputs in a given combination, it is something else to get a predetermined output out of them' (Leibenstein, 1966: 408). The magnitude of the possible improvements in incentive efficiency is represented by an unknown factor X, the reason why Leibenstein introduced the concept under the label of X-efficiency (Frantz, 1992). He suggests that X-efficiency accounts for a great deal of the unexplained residual in economic growth.

Much of the anecdotal evidence on inoptimal energy and resource use in the management of firms cited in support of the Porter hypothesis is similar to the evidence on the use of labour that accumulated in the literature following Leibenstein's hypothesis. There are several good reasons why companies would not be rational and optimal in their use of energy as an input factor, and these reasons go beyond the simple transaction costs of gathering the necessary information and undertaking the required technical changes. They relate to the degree of slack in human behaviour and in company operations, and the failure to mobilize all the knowledge which is embedded in an organization. Energy will be squandered as long as prices are relatively modest compared with other input factors, such as labour and capital, but once outside pressure is introduced, companies will be motivated to mobilize the knowledge and technology available so as to control unit energy costs. Out of such a process, innovations may evolve which may improve economic efficiency and competitiveness.

The literature on the evidence of how energy efficiency can be improved at little or no cost is abundant. In one of the more rigorous explorations, DeCanio (1993: 445) found that energy-saving projects under the US Environmental Protection Agency's (EPA's) Green Lights programme were 'far more profitable than any plausible risk-adjusted cost of capital for comparable projects'. Most of the case studies have failed to bring the results further in order also to assess the benefits at the macroeconomic level (Triebswetter and Hitchens, 2005). However, in revisiting a study on the impact of wastewater taxes on efficiency, Andersen (1999; Andersen *et al.*, 2005) found that in The Netherlands the wastewater sector operates at a lower cost than in countries where a wastewater tax is not employed, and assessed the benefit to be 0.2 per cent of annual GDP.

1.5 Conventional indicators of competitiveness

As noted by Fagerberg (1996), competitiveness is an elusive term. While there are many economic concepts completely unfamiliar to the lay person, when it comes to competitiveness everyone appears to know what it means—and have an opinion on it. If competitiveness was well-defined, our task here could be relatively simple, but there are various understandings and definitions of the concept at play.

There is disagreement on whether it makes sense to talk about 'national competitiveness' in the way that Porter does. Some authors argue that 'competitiveness' applies to firms rather than to countries (cf. the UK's

11

Department of Trade and Industry, 1998,[2] and a much-quoted essay by Krugman, 1994). Thompson (1998) argues that the competitiveness of a country rests on the competitiveness of individual firms, which may not be evenly distributed. One could have a 'leopard spot' economy, with islands of strongly competitive sectors or firms among others which are not. There is no particular reason to expect that all firms and sectors in a country should be at the same level of competitiveness. In any case, most countries have a fairly large domestic sector which is not exposed to competition to the same degree as their export-oriented industries. For this reason, most analysis of competitiveness is focused on the manufacturing sector, leaving services and welfare provision aside.

Despite the ambiguity of the concept of national competitiveness, the OECD and several banking institutions have over many years developed relatively sophisticated indicators for national competitiveness (Durand and Giorno, 1987; Durand *et al.*, 1992; Turner and Van 't dack, 1993). Relative exchange rates are crucial in drawing comparisons between changes in competitiveness, but the problem arises that exchange rates fluctuate differently against different currencies. Furthermore, different competitors are significant in export markets and in the domestic market, respectively. When setting out to compare prices or costs between one or more countries, one would ideally control for changes in exchange rates in a way that adjusts for the relative significance of various markets and competitors in order to draw the right conclusions about changes in competitiveness. As shown by Turner and Van 't dack (1993), however, the problems associated with constructing weighted exchange rates are not trivial. In the case of a real exchange rate, it requires the use of different indices, weighting systems, and specifications of the relationships with a country's trade balance, leading to rather different results depending on the specific approach chosen. One possibility is to take the ratio of one measure to another to paint a wider picture of a country's competitive position, for example, price to cost indices as a proxy for profitability (ibid. 27).

Manufacturing unit labour costs, which reflect salary levels relative to productivity, seem to be the preferred deflator for real exchange rate changes in analysing competitiveness, although often accompanied by other deflators such as export unit costs and consumer prices (Marsh and Tokarick, 1996). The principal disadvantage of using export unit

---

[2] '... the ability to produce the right goods and services of the right quality, at the right price, at the right time. It means meeting customer needs more efficiently and more effectively than other firms' (quoted from Budd and Hirmis, 2004).

costs is that in order to avoid loss of markets, some companies may decide to export products at less profitable or even unprofitable prices. This deflator also ignores competition against imports on the domestic market. The principal disadvantage of using consumer prices is that some goods are under price control and may introduce noise in the calculations. In addition, a significant part of trade is in intermediate goods, and these are not included in consumer price indices. Unit labour costs are based on data that is widely available, also on a comparative basis.

A more radical solution is to calculate absolute levels of competitiveness. Detailed measures of productivity and estimates of purchasing power parities (PPP) have enabled development of level-based measures. This approach is applied by the European Commission, for example, in its Competitiveness Report 2004 (EC, 2004) in a sectoral study of the automotive industry. The principal disadvantage of using PPP is that this is a measure based on domestic expenditure not output; but if GDP is corrected for indirect taxes and imports, this need not be a major issue. However, as the study of the automotive sector shows, there can be differences in the absolute level of unit labour costs, due to, for example, the degree of outsourcing or the amount of intermediate goods supplied, causing difficulties in comparison. Although these types of measures are taken to improve competitiveness, the direct comparison of unit labour costs is somewhat distorted. Especially if the focus is on one or more sectors rather than the manufacturing industry as a whole, such factors tend to amplify differences in an unfortunate way.

In the context of the Porter hypothesis, we should also note that the unit labour cost indicator refers to price competitiveness only, and does not reflect the broader preoccupation, in part of the ETR debate, with incentives for innovation and technological development which may strengthen competitiveness in the longer run.

Indeed, unit labour costs as a short-run indicator require some caution in interpretation (cf. the so-called Kaldor paradox). While increasing unit labour costs normally would be interpreted as loss of competitiveness, they may in fact reflect competitive strength and the ability of a country to market its products as a price-setter, so as to allow for higher labour earnings. Conversely, declining unit labour costs leading to low domestic wages may contradict the OECD definition wherein 'maintaining and expanding the real incomes of its people over the longer term' (OECD, 1992: 237) represents a litmus test of competitiveness.

Perhaps for this reason, some previous analysis of the impact of ETR on competitiveness has gone beyond the traditional indicators and modelled the resulting trends in export market shares (cf. Barker, 1998). Still, due to difficulties with measurement of changes in market shares ex-post, quantitative indicators of real exchange rate value and trend productivity growth as measures of short- and long-term competitiveness respectively are, according to some authors, often regarded as more appropriate (Wagner, 2003: 15).

The context of ETR raises a further difficulty with the use of unit labour costs as an indicator of competitiveness, in that ETR induces differences in energy costs which normally are assumed away. Durand and Giorno (1987) note that raw materials such as energy products are traded at world prices and do not influence relative competitiveness, and as such can be assumed away in comparative analysis. Still, if ETR lowers unit labour costs via increases in energy costs, it seems appropriate also to calculate and consider trends in unit energy costs, as these are altered by ETR. If ETR helps companies focus more attention on improving energy efficiency, one would expect energy unit costs to decline after the initial price shock.

1.6 The need to account for technology and innovation

Productivity growth, reliability, delivery times, quality, after-sales service, financing arrangements, technological innovation, investment in physical and human capital, as well as the institutional and structural environment are all factors that need to be taken into account in assessing the competitiveness of a particular country (Agenor, 1997: 103). But because most of these factors are qualitative, researchers have often abstained from trying to take them into account and have relied mainly on quantifiable indicators, such as unit labour costs.

The paradox is that while conventional indicators related to price or cost competitiveness would predict losses in market shares as a result of increases in the levels of these indicators, the experience in the post-war period, as demonstrated by Fagerberg (1988) and Amendola *et al.* (1993), is that the countries which have experienced the fastest rates of growth in terms of exports and aggregate output have also experienced much higher growth in unit labour costs than other countries.

This 'perverse' relationship (Agenor, 1997) between growth in unit labour costs and growth in export market shares can be explained by taking into account relative technological capabilities. Fagerberg (1988), on the basis of econometric analysis of 15 OECD countries, shows that unit labour costs may play a more modest role than commonly believed. The Kaldor paradox persists, in that increasing unit labour costs and increasing market shares tend to go hand in hand. Instead, it appears that increases in R&D and in productivity correlate better with increases in market shares for exports. Results of other more sector-specific studies suggest that the link between technological activity and export performance is particularly strong in the chemicals and machinery industries, but also exists in less high-tech industries such as metal products, and food and drink (Fagerberg, 1996). Nevertheless, price competitiveness persists in many low-tech, as well as in some high-tech industries. The question, as Agenor (1997) phrased it, remains therefore how to account for the impact of both price and non-price factors.

Addressing non-price factors raises the important question of how these can be identified and measured. One approach, applied by Fagerberg, is to take gross investments in physical capital as a proxy for productive capacity. Another more common approach is to measure both R&D activity and patents. The European Commission in its Competitiveness 2004 report devotes a full chapter to discussion of R&D impacts, and in its study of the automotive industry presents data on the composition of R&D expenditure from the Community Innovation Survey carried out by Eurostat, which allows for a breakdown of innovation expenditure according to various categories and by NACE classification (EC, 2004: 202).

As reflected above, two somewhat different methodological approaches are available for producing assessments of ETR proposals. A *bottom-up* methodology taking specific industries and their opportunities and constraints as the starting point is often favoured by business interests and environmentalists alike, while *top-down* macroeconomic assessments are recommended by economists eager to explore the wider dynamics. While the bottom-up approach leads to inspections of unit labour costs as well as of other above-listed indicators at sectoral level, the top-down approach seeks to integrate partial observations in a wider modelling framework, capturing inter- and intra-sectoral impacts.

In our ex-post assessment of the implications of carbon-energy taxation in the context of ETR, we have opted for the difficult art of combining the two approaches. Macroeconomic modelling in theory should capture the relevant multipliers, but much depends on the properties of the models available and their abilities to fully describe the economic processes (Det Økonomiske Råd 1999). Although we present results from a very comprehensive time-series estimated econometric macroeconomic model, E3ME, with high sectoral and geographic diversity, we felt that scepticism about such models prevails, especially among non-specialists, to the extent that more partial analysis from the bottom-up perspective could be helpful to clarify more transparently some of the partial economic processes at play.

1.7 Coverage of the book

In this book we present the findings of the EU-funded research project COMETR[3] (The Competitiveness Effects of Environmental Tax Reforms). The book is divided into four parts: (1) an introduction explaining how carbon-energy taxes have been introduced as part of ETR in European countries, (2) analysis of the implications for industry-sector competitiveness, (3) analysis of country competitiveness and carbon leakage, and (4) a final section addressing the possible interplay between emissions trading and taxation and outlining the implications of the European experience for wider climate policy-making.

Following this general introduction to the conceptual and theoretical debate, Speck and Jilkova in Chapter 2 provide a detailed overview of carbon-energy taxation and environmental tax reforms in the seven pioneer countries: Denmark, Finland, Germany, The Netherlands, Slovenia, Sweden, and the UK. Their overview summarizes various phases of reforms, unravels revenue recycling mechanisms, and provides detailed data on tax exemptions for energy-intensive industries and other key players. Further reforms have been implemented in new member states, particularly Estonia, which has lowered income taxation by 6 percentage points by phasing in new energy and transport taxes.

1.7.1 Industry-sector competitiveness

As a framework for considering the impacts of carbon-energy taxes, Chapter 3 by Fitz Gerald, Keeney, and Scott explores the fundamental notion of price competitiveness, which has been challenged and qualified

---

[3] COMETR received financial support under the European Union's sixth framework programme for research; see <http://www.dmu.dk/COMETR>.

in the economic literature as referred to above, but which continues to capture the mindsets of many business people and decision-makers who are concerned about day-to-day management, where increases in factor costs have to be matched by improved sales prices and earnings. The key research question which the chapter addresses is the extent to which in particular the energy-intensive industries are price-takers or price-setters; it is only in the latter case that the increased costs associated with carbon-energy taxes can be retrieved with a mark-up on sales prices. If energy-intensive industries are price-takers, unilateral carbon-energy taxes will cause a loss in profitability and ultimately also in competitiveness. As energy-intensive industries are usually not intensive with respect to use of labour, even a revenue-neutral tax reform with recycling of carbon-energy taxes to lower labour costs will not enable a full offset of the changed burden, an issue that the chapter also explores. This asymmetry has been the principal argument for providing tax exemptions for energy-intensive industries, whereby the carbon price signal for these large emitters has been weakened significantly. The chapter by Fitz Gerald *et al.* explores price competitiveness with an econometric panel regression analysis over several decades and finds some diversity among energy-intensive industries; while steel and metal tend to be price-takers, other very energy-intensive sectors, notably non-mineral products, including cement, tend to behave more like price-setters in the market. These findings suggest that exemptions can be granted on a more selective basis than has been the case so far.

In the following three chapters, analysis of the impacts of carbon-energy taxation and ETR proceeds on the basis of a comprehensive database of carbon-energy taxes that has been established as part of our research. Previous research has been largely unable to take account of the complex tax arrangements that are in place for energy-intensive industries and have resorted to use of average energy prices and average carbon-energy tax rates when analysing the ex-ante and ex-post economic implication of ETR. Here we have built a database of sector-specific energy prices and taxes for eight subsectors of energy-intensive industries, collated on the basis of official sources, and make use of it for the purpose of explorative bottom-up analysis as well as for input to the macroeconomic top-down analysis based on the E3ME-model.

Salmons and Miltner in Chapter 4 explore the actual cost increase which carbon-energy taxes have caused for eight energy-intensive sectors in the seven countries in question. As several of these countries are in the Euro-zone, some of the fundamental difficulties with exchange rate fluctuations discussed above can be contained. This analysis shows that although some tax rates have increased by up to 15–20 per cent, the average impact on energy factor costs is much more modest and reaches about 10 per cent only in exceptional cases. However, because of the overall low share of energy costs as a share of total factor costs, the share of increase in production costs is in general minimal—and even for cement reaches only about 2 per cent. The chapter by Salmons and Miltner also explores the overall trends in competitiveness for the 56 sectors in question by considering four traditional indicators, both from the theoretical perspective and from the availability of the sector-specific data collated. Share of global production, import intensity, and to some extent export intensity are measures that provide good indicators of competitiveness. Overall there are no signs of dramatic changes in competitiveness, which leads to the conclusion that more detailed econometric analysis is required to separate out the effects from the relatively small changes in carbon-energy costs.

Such analysis is presented in Chapter 5 by Enevoldsen *et al.*, who use panel regression techniques to separate out the impacts of both energy taxes and energy prices on changes in unit energy costs, and consider their influence ex post, as well as that of unit labour costs. Also this analysis builds on the availability of sector-specific energy prices and taxes for 56 energy-intensive sectors for seven countries that introduced some amount of carbon-energy taxation as part of their environmental tax reforms in the period 1990–2002. The chapter identifies what appears to be a significant difference between the impacts of increased energy prices and increased carbon-energy taxes; whereas increases in energy prices influence unit energy costs, such impacts are considerably smaller for energy taxes. While the analysis does not provide support for a strong double dividend, it cannot confirm the existence of firmly negative economic impacts from ETR.

Andersen and Speck in Chapter 6 review in detail the mitigation and compensation arrangements for energy-intensive industries in the seven countries. Due to significant exemptions for energy-intensive industries, the incentives to improve energy efficiency and shift towards low-carbon fuels for the biggest polluters have been weaker than nominal carbon-energy tax rates would suggest. The analysis in Chapter 6 extends the analysis of sector competitiveness in the two previous chapters by considering the implications of the recycling of revenues. Reductions in taxes on labour paid by employers help reduce the burden of carbon-energy taxation, as do the associated energy efficiency measures.

1.7.2 Country competitiveness and carbon leakage

The macroeconomic analysis with the E3ME model in Chapter 7 by Barker, Junankar, Pollitt, and Summerton addresses competitiveness according to a more comprehensive and dynamic modelling framework. E3ME is a large-scale multi-sectoral integrated energy-environment-economy model of 27 European countries.[4] E3ME models the combined effects of increases in carbon-energy taxes with neutral revenue recycling, either via reduction of social contributions or via lowering of income taxes. The E3ME model ultimately, via changes in import and export ratios, predicts market shares for individual industrial sectors as a result of ETR—and is run ex post, along with a forecast up to 2012 to model the changes in market shares as a result of the actual ETRs in six countries (and also including Slovenia). It is able to capture inter-industry as well as inter-country adaptations to ETR-induced changes in energy prices. The model addresses price competitiveness, but takes account of non-price elements, technological development, and R&D via gross fixed capital formation data. According to E3ME, the ETRs caused both a reduction in fuel use and greenhouse gas emissions and a small increase in employment and GDP. Revenue recycling meant that the cost of ETR to the economy was significantly reduced.

One of the concerns voiced in debates on carbon-energy taxation and ETR as an aspect of competitiveness relates to the risks of so-called 'carbon leakage'. Leakage is said to occur if, as a result of abatement policies, CO_2 emissions are simply displaced to other countries and regions, without such abatement polices. In Chapter 8 Barker *et al.* review the literature on carbon leakage and observe that studies of the effects of the Kyoto Protocol have shown carbon leakage rates (typically from tax and permit schemes with lump-sum revenues recycling) to be in the range of 5–20 per cent, using static computable general equilibrium models. However, in practice, researchers have found that carbon leakage is unlikely to be substantial because transport costs, local market conditions, product variety, and incomplete information all tend to favour local

---

[4] E3ME is a model acknowledged for its ability to support impact assessment in the European Union; see <http://iatools.jrc.ec.europa.eu/bin/view/IQTool/Macro-econometricmodels.html>.

production. Chapter 8 reports findings on carbon leakage from EU member states that implemented ETRs unilaterally over the period 1995–2005.

1.7.3 Implications for future climate policy

The European Union has in place the world's largest carbon emissions trading (the EU Emissions Trading Scheme, or EU ETS), as well as an Energy Taxation Directive which sets minimum rates of excise duties on fossil fuels. Chapter 9 by Ekins explains how these harmonized schemes complement the carbon taxes and energy excise duties which individual member states have in place. The European Commission is currently considering the possibility of splitting its Energy Tax Directive excise duty into energy and carbon components. This proliferation of market-based instruments of energy taxation or carbon control raises a number of very important issues which this chapter explores.

Chapter 10 concludes by placing findings reported in individual chapters in the context of climate policy debates and the time available for stabilizing atmospheric concentrations of CO_2 at a level sufficient to respect the 2 degree target. Pricing of carbon emissions appears to be the instrument of choice, with taxation of carbon a complementary option alongside carbon allowance trading. While the Intergovernmental Panel on Climate Change (IPCC) stipulates that the price of carbon will have to be increased to a level of 20–50 US\$/tCO_2 by 2030 to allow for a trajectory of decarbonizing energy and transport systems, Chapter 10 notes that such price levels are already in place in some countries in Europe. The lessons learnt from the environmental tax reforms in Europe, and the extent to which competitiveness and climate policy can be realigned, are summarized with a view to their potential relevance for future policies considered in Europe as well as by other major emitters.

References

Agenor, P. R. 1997. 'Competitiveness and external trade performance of the French manufacturing industry'. *Weltwirtschaftliches Archiv*, 133/1: 103–33.

Amendola, G., Dosi, G., and Papagni, E. 1993. 'The dynamics of international competitiveness'. *Weltwirtschaftliches Archiv*, 129/3: 451–71.

Andersen, M. S. 1999. 'Governance by green taxes: implementing clean water policies in Europe'. *Environmental Economics and Policy Studies*, 2/1: 39–63.

——Lerche, D. B., Kristensen, P., and Smith, C. 2005. 'Effectiveness of urban waste-water policies'. EEA report 2. Copenhagen: European Environment Agency.

Ashford, N., Ayers, C., and Stone, R. 1985. 'Using regulation to change the market for innovation'. *Harvard Environmental Review*, 9/2: 419–65.

Barker, T. 1998. 'The effects on competitiveness of coordinated versus uni-lateral fiscal policies reducing GHG emissions in the EU: an assessment of a 10% reduction by 2010 using the E3ME model'. *Energy Policy*, 26/14: 1083–98.

——and Köhler, J. (eds.) 1998. *International Competitiveness and Environmental Policies*. Cheltenham: Edward Elgar.

Bovenberg, A. L., and de Mooij, R. A. 1994. 'Environmental Levies and Distor-tionary Taxation'. *American Economic Review*, 84/4: 1085–9.

——— 1997. 'Environmental Tax Reform and Endogenous Growth'. *Journal of Public Economics*, 63: 207–37.

Bruvoll, A., and Larsen, B. M. 2004. 'Greenhouse gas emissions in Norway: do carbon taxes work?' *Energy Policy*, 32: 493–505.

Budd, L., and Hirmis, A. K. 2004. 'Conceptual framework for regional competitive-ness'. *Regional Studies*, 38/9: 1015–28.

Commission of the European Communities. 1993. 'Growth, competitive-ness, employment: the challenges and ways forward in the 21st century'. COM/93/700, 5.12.1993.

Department of Trade and Industry (UK) 1998. Regional competitiveness indicators. London: HMSO.

DeCanio, S. 1993. 'The efficiency paradox: bureaucratic and organizational barriers to profitable energy-saving investments'. *Energy Policy*, 26/5: 441–54.

Det Økonomiske Råd. 1999. 'CO$_2$-skatter, dobbelt-dividende og konkurrence i energi-sektoren: Anvendelser af den danske AGL-model ECOSMEC'. *Arbejdspapir*, 1999/1, København.

Duff, D. 2009. 'Carbon Taxation in British Columbia'. *Vermont Journal of Environ-mental Law*, 10: 85–105.

Durand, M., and Giorno, C. 1987. 'Indicators of international competitiveness: conceptual aspects and evaluation'. Paris: OECD.

——Simon, J., and Webb, C. 1992. 'OECD's indicators of international trade and competitiveness'. Working Papers 120, GD(92)138. OECD: Economics Department.

Enevoldsen, M. 2005. *The Theory of Environmental Agreements and Taxes*. Chel-tenham: Edward Elgar.

Esty, D., and Porter, M. 2001. 'Ranking national environmental regulation and performance: a leading indicator of future competitiveness'. *The Global Compet-itiveness Report 2001–2002*. Oxford: Oxford University Press.

European Commission. 2004. *European Competitiveness Report*. SEC(2004)1397. Brussels.

Fagerberg, J. 1988. 'International competitiveness'. *Economic Journal*, 98: 355–74.

Fagerberg, J. 1996. 'Technology and competitiveness'. *Oxford Review of Economic Policy*, 12/3: 39–51.

Frantz, R. 1992. 'X-efficiency and allocative efficiency: what have we learned?' *American Economic Review*, 82/2: 434–8.

Goodstein, E. 2003. 'The death of the Pigovian tax? Policy implications from the Double-Dividend debate'. *Land Economics*, 79/3: 402–14.

Goulder, L. 1995 'Environmental Taxation and the "Double Dividend": A Reader's Guide'. *International Tax and Public Finance*, 2/2: 157–83.

Jaffe, A., Peterson, S., and Stavins, R. 1995. 'Environmental regulation and the competitiveness of US manufacturing: what does the evidence tell us?' *Journal of Economic Literature*, 33: 132–63.

Krugman, P. 1994. 'Competitiveness: A Dangerous Obsession'. *Foreign Affairs*, 73/2: 28–44.

Leibenstein, H. 1966. 'Allocative efficiency vs. "X-efficiency" '. *American Economic Review*, 56/3: 392–415.

McCoy, D. 1997. 'Reflections on the double dividend debate', in T. O'Riordan (ed.), *Ecotaxation*. London: Earthscan, 201–14.

Marsh, I. W., and Tokarick, S. P. 1996. 'An assessment of three measures of competitiveness'. *Weltwirtschaftliches Archiv*, 132/4: 700–22.

Nielsen, S. B., Pedersen, L. H., and Sørensen, P. B. 1995. 'Environmental policy, pollution, unemployment and endogenous growth'. *International Tax and Public Finance*, 2: 185–205.

Nutzinger, H. G., and Zahrnt, A. Z. 1990. *Für eine ökologische Steuerreform*. Frankfurt a.M.: Fischer Verlag.

OECD. 1993. *Environmental Policies and Industrial Competitiveness*. Paris: OECD.

—— 1997. *Environmental Taxes and Green Tax Reform*. Paris: OECD.

O'Riordan, T., ed. 1997. *Ecotaxation*. London: Earthscan.

Palmer, K., Oates, W., and Portney, P. 1995. 'Tightening environmental standards: the benefit-cost paradigm or the no-cost paradigm?' *Journal of Economic Perspectives*, 9/4: 119–32.

Parry, I. 1995. 'Pollution taxes and revenue recycling'. *Journal of Environmental Economics and Management*, 29: 64–77.

Pearce, D. 1991. 'The role of carbon taxes in adjusting to global warming'. *Economic Journal*, 101: 938–48.

Porter, M. 1990. *The Competitive Advantage of Nations*. New York: The Free Press.

—— 1991. 'America's green strategy'. *Scientific American*, 264: 168.

—— 1998. *The Competitive Advantage of Nations*, 2nd edn. New York: The Free Press.

—— and van der Linde, C. 1995. 'Toward a new conception of the environment-competitiveness relationship'. *Journal of Economic Perspectives*, 9/4: 97–118.

Rapanos, V. T., and Polemis, M. L. 2005. 'Energy demand and environmental taxes: the case of Greece'. *Energy Policy*, 33: 1781–8.

Smith, J. B., and Sims, W. A. 1985. 'The impact of pollution charges on productivity growth in Canadian brewing'. *Rand Journal of Economics*, 18/3: 410–23.

Smith, S. 2003. 'Environmental taxes and competitiveness: an overview of issues, policy options and research needs'. COM/ENV/EPOC/DAFFE/CFA(2001)90/FINAL. Paris: OECD.

Thompson, G. 1998. 'International competitiveness and globalization: frameworks for analysis, connections and critiques', in T. Barker and J. Köhler (eds.), *International Competitiveness and Environmental Policies*. Cheltenham: Edward Elgar, 13–32.

Triebswetter, U., and Hitchens, D. 2005. 'The impact of environmental regulation on competitiveness in the German manufacturing industry: a comparison with other countries of the European Union'. *Journal of Cleaner Production*, 13: 733–45.

Turner, P., and Van 't dack, J. 1993. 'Measuring international price and cost competitiveness'. Economic Paper 39. Bank for International Settlements, Basel.

van der Linde, C. 1993. 'The micro-economic implications of environmental regulation: a preliminary framework', in OECD, *Environmental Policies and Industrial Competitiveness*. Paris: OECD, 69–77.

Vehmas, J. 2005. 'Energy-related taxation as an environmental policy tool: the Finnish experience 1990–2003'. *Energy Policy*, 33: 2175–82.

von Weizsäcker, E. U., and Jesinghaus, J. 1992. *Ecological Tax Reform*. London: Zed Books.

Wagner, M. 2003. 'The Porter hypothesis revisited: a literature review of theoretical models and empirical tests'. Centre for Sustainability Management, Universität Lüneburg.

Weinbrenner, D. 1999. *Ökologischen Steuerreform: Wirkungszusammenhänge zwischen Emissions- und Fiskal Steuern*. Wiesbaden: Deutscher Universitätsverlag.

2

Design of Environmental Tax Reforms in Europe

Stefan Speck[1] and Jirina Jilkova[2]

2.1 Introduction

Significant tax reforms have been undertaken in European countries during the 1990s, their main objectives being to reduce labour costs and broaden the overall tax base, with an increase in general consumption taxes, such as VAT and environmental taxes. Nordic governments were the first during this period to adopt the concept of environmental tax reform (ETR), followed in the late 1990s by the governments of The Netherlands, UK, and Germany. The strategy followed by these countries was to launch new environmental taxes, in the majority of cases levied on energy consumption and on CO_2 emissions, and to revise already existing environmental taxes. The other component of the tax shift programmes was mainly to reduce income taxes and non-wage costs, such as social security contributions, and to raise personal income tax allowances.

A number of the tax reform packages also included components that directly addressed the business sector, providing support schemes for investment in energy efficiency. Only recently have tax reform proposals to reduce the capital tax burden of industry gained prominence on the political agenda; in particular, as a consequence of the lower capital tax rates levied in some of the new EU member states.

[1] Stefan Speck, Senior Economist, Department of Policy Analysis, National Environmental Research Institute, Aarhus University, Denmark and Kommunalkredit Public Consulting, Austria.
[2] Jirina Jilkova, Professor, Economics University of Prague, Czech Republic.

Revenue generated from environmental taxes expressed as a share of total tax revenue or GDP is regularly used as an indicator to illustrate the significance of environmental policy in a country. Interpretation of such comparisons must be treated with some caution, because the figures do not say anything about the environmental appropriateness of the overall fiscal policy or environmental policy in the country. Nevertheless, comparisons of this kind, for example, comparison of the trend in the labour-taxation-to-GDP ratio with the environmental-tax-to-GDP ratio, are regularly used to study whether a country is moving in the direction of ETR (EEA, 2005).

Throughout the 1990s, the labour-taxation-to-GDP ratio increased in many EU-15 member states, but the rate of increase slowed from the mid-1990s to 2005. During the 1990s, the environmental-tax-to-GDP ratio also increased in the majority of EU member states as a consequence of the increased use of environmental taxes. During recent years, however, the environmental-tax-to-GDP ratio has been more or less constant in the majority of EU member states, with some exceptions—in Denmark the ratio peaked in 2005 and in the UK the ratio moved in the other direction, falling from 3.1 per cent in 2000 to 2.5 per cent in 2005 (Table 2.1).

Table 2.1. Analysis of taxes on labour vs. environmental taxes between 1990 and 2005 in member states that implemented ETR during this period

	1990	1995	2000	2005
Finland				
Labour tax as % of GDP	24.8	26.1	23.7	23.3
Env tax as % of GDP	2.2	2.9	3.1	3.0
Denmark				
Labour tax as % of GDP	24.1	28	26.6	24.8
Env tax as % of GDP	3.6	4.4	5.2	5.8
Sweden				
Labour tax as % of GDP	35.8	31	32.3	31.2
Env tax as % of GDP	3.4	2.8	2.8	2.9
Germany				
Labour tax as % of GDP	20.9	24.9	24.3	22.3
Env tax as % of GDP	2.0	2.4	2.4	2.5
Netherlands				
Labour tax as % of GDP	25.8	22.1	20.3	17.7
Env tax as % of GDP	3.1	3.5	3.9	4.0
UK				
Labour tax as % of GDP	14.3	14	14.3	14.4
Env tax as % of GDP	2.7	2.9	3.1	2.5

Source: Eurostat.

In all the member states that implemented ETR between 1990 and 2005 (Denmark, Finland, Germany, The Netherlands, Sweden, and the UK), the anticipated result of ETR—a decrease in the labour-taxation-to-GDP ratio and an increase in the environmental-tax-to-GDP ratio—materialized, except in the case of the UK. Here, the labour-taxation-to-GDP ratio remained almost constant throughout the period and the environmental-tax-to-GDP ratio increased throughout the 1990s but dropped between 2000 and 2005.

As in the 'old' EU-15 member states, in the new EU member states the major part of environmental tax revenues are generated via taxes levied on energy products (between 65 and 80 per cent), with the exception of Cyprus and Malta, where around 60 per cent of environmental tax revenue can be attributed to transport-related taxes. In the near future, a number of increases in energy tax revenues will occur in the new EU member states as a consequence of the adoption of the Energy Taxation Directive (Directive 2003/96/EC of 27 October 2003, EC 2003). This directive, restructuring the Community framework for the taxation of energy products, clearly widens the scope of the former EU energy taxation framework under the Mineral Oils Directive (Directive 1992/82/EEC), which set minimum excise tax rates for mineral oil products only. The new framework for the taxation of energy products extends the number of energy products, as minimum tax rates are set for all energy products, including natural gas, coal, and electricity, as well as increasing the minimum rates for transport fuels. All EU member states, including the new EU member states, are obliged to comply with fiscal structures and the levels of taxation laid down in the 2003 Energy Taxation Directive. However, the new EU member states negotiated temporary exemptions and transitional periods for full compliance to avoid potential, serious, economic and social difficulties arising from direct transposition during economic transition. The transitional periods granted to the new EU member states—in some cases up to 2009/2010—will reduce the initial revenue-generating effect of energy taxes. In addition, some of the new member states already impose tax rates on transport fuels (petrol and diesel) that exceed EU minimum rates—therefore here there has been no need for further increases. This is important whendiscussing ETR, as taxes levied on these energy products generate by far the largest revenue streams.

The following sections describe the major ETRs implemented in EU member states and, in addition, illustrate the development of the energy and carbon taxation scheme in one of the new EU member states, namely Slovenia. This country is chosen because it was the first of the ten new

EU member states to introduce a CO_2 tax with the aim of lowering greenhouse gas emissions and reducing consumption of non-renewable energy products (REC, 1999). One focus is a detailed discussion of the special tax provisions that have been implemented in order to address the risk, often quoted by industry, of losing international competitiveness when a country introduces carbon/energy taxes unilaterally. Our discussion centres around carbon/energy taxes levied on energy products used for industrial/commercial purposes and heating and not on transport fuels.

2.2 Denmark

Denmark was one of the first countries in Europe to introduce a CO_2 tax. The CO_2 tax represented an addition to the energy taxes already in place on oil products, coal, and electricity consumption. The CO_2 tax was introduced in two phases: in May 1992 it was applied to energy products consumed by households and in January 1993 it was extended to businesses.[3] It was moreover accompanied by a reduction in the rates of existing energy taxes.

Introduction of the CO_2 tax represented a turning point in energy taxation in Denmark, especially for industrial energy consumption, because until this point industry had been exempt from energy taxation. During the period 1993 to 1995, industry was granted a 50 per cent reduction in the CO_2 tax rate; therefore, instead of being subject to approximately 13 EUR (100 DKK) per tonne of CO_2 energy consumption, industry was subject to a rate of 6.6 EUR (50 DKK) per tonne CO_2. A three-tiered reimbursement scheme granting further tax relief according to the energy intensity of each business was put in place in 1993. The refund scheme was differentiated based on actual energy costs paid and in relation to total sales (Malaska *et al.*, 1997):

- If the CO_2 tax burden was between 1 and 2 per cent of the difference between sales and purchases (i.e. net sales), the company was eligible for a tax refund of 50 per cent of the sum exceeding the 1 per cent limit.
- If the CO_2 tax burden was between 2 and 3 per cent of the difference, the tax refund amounted to 75 per cent of the sum exceeding the 2 per cent limit.

[3] Industry was also liable to pay CO_2 tax in 1992, but in 1992 industry received a refund of the entire CO_2 tax paid—see: Nordic Council of Ministers (2006).

- If the CO_2 tax burden was above 3 per cent of the difference, the tax refund was 90 per cent of the sum exceeding the 3 per cent limit. Companies falling under the 90 per cent refund scheme could receive additional tax support covering the remaining part of the CO_2 tax burden. However, this support was limited to three years only and the company had to pay at least 10,000 DKK (1,320 EUR) in CO_2 tax. This refund scheme was in place until 1995[4] and was revised when the second phase of the Danish ETR was introduced in 1996.

Three phases of the Danish ETR reform process can be distinguished. The first phase, the 1993 ETR, covered the period 1994–8 and targeted mainly the household sector:

- The political objective of the ETR was to reduce the marginal tax rates levied on personal income (Jensen, 2001).
- The programme of tax shifts: The revenue losses following the income tax rate reduction amounting to approx. 2.3 per cent of GDP in 1998 were partly offset by increased revenues from environmental taxes amounting to an expected 1.2 per cent of GDP and payroll taxes amounting to around 1 per cent of GDP.
- Additional environmental taxes were introduced; namely a tax on tap water, a wastewater tax, and a tax on plastic and paper bags. Revenues from increased energy taxes accounted for 7.5 billion DKK (1 billion EUR) of the projected 12 billion DKK (1.6 billion EUR) generated from environmental taxes.

The 1995 tax reform, that is, phase 2, was implemented during the period 1996–2000 and the main sector affected was industry.

- The size of the tax shift programme was smaller than in the first phase as revenues generated from environmental taxes were projected to amount to 2.45 billion DKK (330 million EUR)—approx. 0.2 per cent of GDP in 2000.
- The programme of tax shifts: The revenues raised by increasing the energy tax rates, as well as the introduction of a sulphur tax and an energy tax on natural gas, were used to reduce employers' social security contributions and to provide subsidies for investment in energy efficiency programmes. The main recycling mechanisms adopted in relation to industry have been:

[4] This additional refund scheme reduced the average CO_2 tax burden to around 35 per cent of the standard rate, i.e. a rate of 4.6 EUR (35 DKK) per tonne CO_2.

(i) provision of investment grants for energy-saving measures;
(ii) recycling of a fraction of the revenues to private enterprises, comprising two elements:

- a reduction in employers' contributions to the additional labour market pension fund amounting in 1996 to 1,325 DKK (177 EUR) per year, per employee, compared to 1,166 DKK (156 EUR) in 1995;
- a reduction in employers' national insurance contributions according to the Act on Labour Market Funds: contributions to be lowered by 0.11 percentage points in 1997, 0.27 in 1998, 0.32 in 1999, and 0.53 in 2000;

(iii) establishment of a special fund for small and medium-sized enterprises, as it was expected that these enterprises would only benefit in a small measure from (ii).

An overview of the expected revenues generated by industry and households as well as the recycling mechanisms is presented in Table 2.2.

Additionally, in phase 2 the industrial energy taxation scheme was overhauled. Industrial energy consumption was subdivided into three components: space heating, light processes, and heavy processes. The rationale behind this reform was that industry should, in part, face the same energy tax rates as households. Industry had the same tax burden

Table 2.2. Phase II of Danish ETR, 1996–2000

	1996	1997	1998	1999	2000
Total tax revenue collected	915	1,440	1,955	2,220	2,450
Industrial and commercial	710	1,230	1,730	1,900	2,075
—Space heating	420	750	1,050	955	910
—CO_2 tax	65	245	425	585	775
—SO_2 tax	225	235	255	360	390
Households	205	210	225	320	375
Revenue recycled					
To trade and industry	710	1,230	1,730	1,900	2,075
—Investment subsidies	300	500	500	500	0
—Small businesses	180	210	255	255	295
—Reductions in employers' SSC	200	490	945	1,115	1,750
Administration costs	30	30	30	30	30
Compensation to households	145	150	165	240	315
*To electric heating users**	60	60	60	60	60

Notes: *Subsidies for conversion to electric heating; revenue figures planned as of 1995 in million DKK.
Source: Hansen (1999).

as households for energy used for space heating; that is, industry paid the full energy tax as well as the full CO_2 tax. However, energy used for activities other than space heating was still fully exempt from energy tax and a reduced CO_2 tax rate applied, differentiated according to actual purpose.

The third phase was the 1998 tax reform implemented during the period 1999–2002.

- The tax shift programme was planned to be in the range of around 6.4 billion DKK (850 million EUR) over the period 1999–2002, amounting to approximately 0.3 per cent of GDP in 2002.
- The programme of tax shifts: Increased revenues from environmental taxes as well as corporate taxes were used to reduce personal income tax rates and taxes levied on the yield of pension savings and share yields.
- The revenues for the recycling programme were raised via increases in energy taxes. During this period the tax on petrol, light fuel oil, and heavy fuel oil was raised by 5–7 per cent, the tax on diesel by 16 per cent, coal by 12 per cent, electricity by 15 per cent, and natural gas by 33 per cent.

It is noteworthy that the recycling mechanisms implemented clearly reflected the contribution of the two different economic sectors. Industry and households received the amount which they were expected to be paying as a consequence of the reform process, that is, no cross-subsidization of any type was to occur. This Danish approach of a fair and equal distribution of revenues has to be seen and compared with the German approach as discussed below. Moreover, the personal income tax reduction in Denmark mainly affected those with lower and medium incomes and compensation for pensioners was also included. As mentioned above, the main revenue raising policy was to increase energy tax rates and not CO_2 tax rates. This is in itself significant, however, because the industrial sector is not greatly affected by energy tax increases due to the special tax provisions that apply.

Special tax provisions for industry

Industry has benefited from favourable energy tax provisions since the 1990s, although these special tax provisions have changed over time. As mentioned above, the sector was not subject to any taxes levied on energy products until the introduction of the CO_2 tax in 1992. Between 1993 and

1995 enterprises were subject to 50 per cent of the standard CO_2 tax and in addition energy-intensive industries were eligible for a special CO_2 tax refund scheme according to the CO_2 tax liability measured with respect to value added (Nordic Council of Ministers, 1994).

The scheme of special tax provisions for industry changed in 1995. As mentioned above, companies paid CO_2 taxes which varied according to usage—the full CO_2 tax rates applied to space heating, and process purposes, differentiating between heavy and light processes, were generally exempt from any energy taxation. This still applies today.

Energy consumed in processes other than space heating was generally levied with a CO_2 tax rate which increased gradually from 50 DKK (6.7 EUR) per tonne CO_2 up to 90 DKK (12.1 EUR) per tonne CO_2 in 2000. Companies entering an agreement with the Danish energy authority were eligible for a reduction in the CO_2 tax rate on increasing their energy efficiency. The Danish government used a list[5] of energy-intensive processes categorized as heavy processes to define whether enterprises could be classified as energy intensive and whether the CO_2 tax rate could therefore be further reduced. They also applied the so-called 'Proms criterion'— whereby the financial strain posed by taxes in relation to the value added generated in the enterprise was used. Based on this, an enterprise was classified to be energy intensive if the liability incurred from a tax of 50 DKK per tonne CO_2 was to permanently exceed 3 per cent of value added in the enterprise, while the liability simultaneously exceeds 1 per cent of sales.

It is interesting to note that the standard CO_2 tax rate of 100 DKK (13.4 EUR) per tonne CO_2 has not been increased since the tax was implemented in 1992. The CO_2 tax burden of industry increased gradually during the period of the second phase of the ETR (1996–2000) but then remained constant until 2004. In 2005 the nominal CO_2 tax rate of 100 DKK was reduced to 90 DKK (12.1 EUR) per tonne CO_2. However, this revision does not affect the effective tax rates paid by industry, as the share was increased proportionally so that the tax rate per tonne CO_2 remained the same as before (Nordic Council of Ministers 2006). This reduction in the CO_2 tax rate was compensated for with a slight increase in energy tax rates, but this does not affect industry because this sector is still exempt from energy tax, apart from on energy used for space heating. Although effective tax rates are rather low, especially when compared to the high Danish nominal tax rates, they are still in excess of the minimum excise

---

[5] The European Commission approved the process list as part of the overall Danish energy taxation scheme.

tax rates set by the European Union in the Energy Taxation Directive (Directive 2003/96).

Electricity consumption in manufacturing industry has also been subject to a number of special tax provisions. The standard Danish electricity tax, comprising an energy tax component and a CO_2 tax component, is one of the highest electricity taxes in Europe. The scheme applies a two-tier approach, distinguishing between electricity used for heating purposes and for all other uses. Furthermore, manufacturing industry enjoys some further tax provisions for consumption of electricity for process purposes by virtue of some favourable exemption rules. First of all, in 2004 companies paid an energy tax rate of 10 DKK (1.3 EUR) per MWh, as compared to the standard rate of 566 DKK (76 EUR) per MWh, and they were also eligible for a 40 per cent refund of the CO_2 tax of 100 DKK (13.4 EUR) per MWh. Therefore, in 2004, the electricity tax rate paid by manufacturing industry amounted to 70 DKK (9.4 EUR) per MWh as compared to 566 DKK per MWh, that is, an effective tax rate of around 12 per cent of the nominal tax rate. Furthermore, the reduced energy tax rate only applied to the first 15 million kWh consumed each year (Nordic Council of Ministers, 2006). Danish energy and carbon taxes were increased throughout the 1990s, but this changed at the beginning of the 2000s. The nominal tax rates were frozen during the period 2002 to 2007 and only in 2008 can a slight increase be reported. This policy has led to a reduction in the real value of energy and CO_2 tax rates. There are further plans to increase the CO_2 tax rate from its current level of 90 DKK (12.1 EUR) per tonne CO_2 to 150 DKK (20 EUR) per tonne CO_2 during 2008, offset by a corresponding reduction in energy tax rates.

2.3 Finland

Finland, in 1990, was the first country in Europe to introduce a CO_2 tax; the tax was levied on all energy products with the exception of transport fuels. At the time, transport fuels were already subject to energy taxes in Finland.[6] The Finnish CO_2 tax must therefore be perceived as an additional tax on energy products, just as in the other EU member states discussed in this chapter.

The design of the CO_2 tax has changed several times during its implementation:

[6] Other oil products were also subject to a form of environmental tax in the form of an oil pollution fee and a precautionary stock fee (Nordic Council of Ministers, 2006).

- Between 1990 and 1994, the CO_2 tax was based solely on the carbon content of the energy product.
- From 1994 up to 1996, the design changed and the CO_2 tax was based on the carbon content as well as on the energy content of the energy product; at the beginning 60 per cent of the tax was determined by the carbon content and 40 per cent by the energy content. This ratio changed to 75:25 during this time period.
- In 1997, the design changed once again and since this time the CO_2 tax has been a pure carbon dioxide tax.

In 1990, the CO_2 tax rate was set at approximately 1.2 EUR per tonne CO_2 and since then has been regularly increased, reaching approximately 18 EUR tonne CO_2 in 2003 and 20 EUR per tonne CO_2 in 2008. The situation in Finland can be compared with that in Denmark, as in both countries tax rates have been kept constant for a while. In Finland, rates have increased only in the last two years, for example, in 2008 the energy tax has increased by 9.8 per cent on average, and the CO_2 tax by 13 per cent.

The Finnish ETR reform process can be divided into two phases (Hoerner and Bosquet, 2001). The first phase started in 1997. From the beginning, this programme of tax shifts was not planned to be revenue neutral and the motivation of the programme was to reduce general tax revenues by 5.5 billion FIM (around 935 million EUR and 0.9 per cent of GDP). The two main components of the programme were:

- A reduction in state personal income tax amounting to 3.5 billion FIM (590 million EUR) and a reduction in employers' social security contributions and in local personal income tax of around 2 billion FIM (340 million EUR).
- This shortfall in revenue was partly compensated for by revenues generated from the CO_2 tax and the landfill tax. The revenues summed up to around 1.4 (1.1 and 0.3, respectively) billion FIM (190 million and 50 million EUR); that is, 0.2 per cent of GDP.

The second phase was agreed to in late 1997 and implemented in 1998. Again the policy did not aim to be revenue neutral. The political objective of this programme was to further reduce labour taxes and to offset some of the deficit by increases in environmental taxes and corporate profit tax, broadening the tax base. The reduction in labour taxation was planned to be:

- 1.5 billion FIM (250 million EUR) in 1998;
- 3.5 billion FIM (590 million EUR) in 1999 (amounting to approx. 0.5 per cent of GDP).

The reform process anticipated a deficit in both years, that is, the reform was never planned to be revenue neutral. The underlying assumption of this policy of reducing taxes levied on a factor of production, labour, was that it would lead to an increase in employment followed by an increase in labour-related tax revenues.

The Finnish ETR affects both households and industry, although the recycling measures favour the household sector. Furthermore, the Finnish energy taxation scheme does not distinguish between different economic sectors, with the exception of electricity taxes, which discriminate between households and industry in the sense that industry faces a lower tax rate (Speck, 2007a). Special tax provisions, such as reductions in energy tax rates and/or complete exemptions from carbon and energy taxes, which in neighbouring Nordic countries have largely been the rule, have never been implemented in Finnish energy policy. The only favourable provisions are lower electricity taxes for industry and that energy products used as raw materials in the production process are exempt from electricity taxes. This policy is not unusual in terms of international energy policies and also accords with EU energy policy. However, when assessing the lack of special provisions to industry in the Finnish energy taxation scheme, the fact that nominal energy tax rates in Finland are generally lower than nominal rates implemented in neighbouring countries should be taken into consideration. However, energy-intensive industries are eligible for a refund mechanism, first implemented in 1998, that relieves these industries from a part of their energy tax burden. The indicator assessing whether a company is energy intensive is different from that used in Denmark, as the Finnish indicator explicitly defines a company as energy intensive if the energy excise taxes paid amount to 3.7 per cent of the value added of the enterprise, not taking into account taxes levied on motor fuels and any tax subsidies received. The companies that qualify are eligible for a tax refund of 85 per cent of energy taxes paid—but the refund only applies if the tax liability exceeds 300,000 FIM (around 51,000 EUR). In 1999, twelve companies, mainly in the paper and pulp industry, were able to receive some reimbursement under this refund scheme and a total of 85 million FIM (14.3 million EUR) was refunded (Nordic Council of Ministers, 2006).

2.4 Germany

The German energy taxation scheme relies heavily on taxes levied on transport fuels, as is the case in all other European countries. However, some differences between Germany and the Nordic countries can be recorded as, for example, coal was not subject to an energy tax in Germany until two years ago, when a tax on coal and coke was introduced. Germany introduced an energy tax on natural gas in 1989. However, it is worth mentioning the electricity taxation scheme that was in operation until 1995. This tax scheme was known under the term 'Kohlepfennig'. The revenues of this tax were earmarked for subsidization of the German coal industry and the scheme was abolished in 1995. The energy tax scheme experienced some major changes in connection with ETR, which was finally implemented in 1999, although the first discussions concerning the potential of ETR started in the early 1990s.

The German ETR

The German government had two objectives when it introduced the ETR in 1999:

- improved environmental protection and in particular reduced greenhouse gas emissions as a means to address climate change;
- reduced employers' and employees' statutory pension contributions in order to reduce labour costs and to increase employment.

The ETR was implemented in two phases. The first phase covers the period between 1999 and 2003 and the main policies were an increase in existing energy taxes and introduction of an electricity tax:

- mineral oil taxes on transport fuels (petrol and diesel) were gradually and steadily increased in five steps between April 1999 and 2003;
- taxes on natural gas and light heating fuels were increased in 1999 and, for natural gas, again in 2003;
- taxes on heavy fuel oil were increased in 2000 and again in 2003. However, the tax on heavy fuel oil used for electricity generation was reduced in 2000 to make the tax uniform across different uses;
- introduction of an electricity tax in 1999, which was gradually increased in five annual steps;

- the increases in the tax rates for energy products, other than transport fuels, imposed on the manufacturing industry and the agricultural sector were lower than the standard increases because of the fear of negatively affecting the competitiveness of German industry.

The ETR was planned to be revenue neutral. But the German government diverged from this policy goal by using a small fraction of revenues to consolidate the federal budget, but only as a temporary measure. The major share of the revenue was used in a programme of tax shifts in which employers' and employees' social security contributions (public pension contributions) were reduced equally. Furthermore, a very small fraction was earmarked for promotion of renewable energy. The total volume of the tax shift programme was 18.6 billion EUR in 2003 (approx. 0.9 per cent of GDP). The recycling mechanism adopted resulted in an estimated 1.8 per cent reduction in employers' and employees' pension contributions from 20.3 per cent in 1998 to 19.5 per cent in 2003. It is estimated that without the introduction of the ETR, the total pension contribution would have been in the region of 21.2 per cent in 2003 as a consequence of economic and demographic development in Germany.

One point of interest in the German ETR approach is a slight inconsistency within the programme of tax shifts. The main economic sector affected by the revised energy tax schemes was the household sector. But the main beneficiary of the imposed recycling mechanisms was the industrial sector as a whole (Bach, 2004).

The second phase of the ETR commenced in 2004. The German government wanted to extend the ETR to a wider Environmental Fiscal Reform (EFR) by focusing on reducing environmentally harmful subsidies and other tax reductions and by adapting the heating fuel tax on natural gas and on heavy fuel oil. The latter policies have been undertaken, but the reform idea of carrying out a major shake-up with regard to subsidies was abandoned because of political resistance, in particular from the opposition parties.

The total revenue raised by energy taxes under the umbrella of the ETR amounted to around 18.6 billion EUR in 2003. The largest part of the revenue was generated by the tax on petrol and diesel (approx. 10.3 billion EUR) and by the newly introduced tax on electricity, amounting to 6.5 billion EUR. It is interesting to note that all energy tax rates have been frozen since 2003, meaning that their real value has fallen. Furthermore, consumption of transport fuel fell in Germany by about 13 per cent between 1998 and 2007 which, combined with the fact that energy tax rates have remained constant since 2003, means that the revenues dedicated to reducing social security contributions have fallen too (Speck, 2007b).

The German ETR complied with the underlying principle of ETR, namely a shift in the tax burden from labour to energy use; however, a qualification to this is that the shift still preferentially benefits industry. This favourable treatment of industries becomes more obvious when the special provisions with regard to the industrial energy taxation scheme are examined.

Special tax provisions for industry

Manufacturing industry as well as agriculture, forestry, and fisheries have been granted special energy tax provisions from the beginning of the ETR. The sectors were granted tax relief of 80 per cent for energy products other than transport fuels. The tax relief affected only the tax rates which were imposed as part of the ETR and were only available on condition that the base sum (*Sockelbelastung*) of 512.50 EUR per annum spent on electricity and heating fuels was exceeded. The manufacturing sector was eligible for an additional tax option—companies could apply for a tax cap (*Spitzenausgleich*). If the tax burden from increased energy tax rates was 20 per cent higher than the tax relief obtained in connection with pension contribution reductions, companies were refunded the full difference. The outcome of this provision was that some industries had an effective tax rate of zero per cent.

A slight revision to the scheme was implemented in 2003, as tax relief for the manufacturing sector, agriculture, forestry, and fisheries was reduced to 40 per cent of the standard energy tax rates for electricity, heating oil, and natural gas, and only for energy consumption exceeding the base sum (*Sockelbelastung*), which was kept constant. The relief only applied to that part of the energy taxes that resulted from the ETR, that is, the tax rate increases which took place after 1999. This policy change means that the effective tax rate is now 60 per cent of the standard rate as compared to only 20 per cent during the period 1999–2003. Moreover, the *Spitzenausgleich* still applies to manufacturing industry but in a slightly different form. The revised rule stipulates that a company is eligible for a refund if the energy tax burden is greater than the tax relief obtained in connection with pension contribution reductions. The refund currently amounts only to 95 per cent of the difference. The outcome of this revision is that companies which receive the tax refund are now subject to a tax rate of 3 per cent as compared to zero per cent under the 1999 regulation. As an example of how this works out, the standard electricity tax rate in 2004 was 20.5 EUR/MWh. Companies which are statistically classified as manufacturing, agriculture, forestry, or fisheries are subject to an effective tax rate of 60 per cent of the standard rate, that is, the electricity tax rate amounts to 12.3 EUR/MWh. Manufacturing companies are subject to an even lower effective tax rate of 0.62 EUR/MWh when they qualify for the *Spitzenausgleich* regulations. During recent years, this rule has been further weakened in the sense that the 60 per cent rule applies to the total energy tax rate.

Further special tax regulations do exist, but they are not specifically directed to addressing the competitiveness concerns of industry. These provisions are mainly designed for environmental purposes, either to promote renewable energies or to promote improvements in energy efficiency. For example, highly efficient CHP facilities with a monthly or annual utilization rate of 70 per cent or more are fully exempt from the mineral oil tax.

As seen above, the German system of energy taxation, in particular that introduced in the 1999 ETR process, includes a whole range of special tax provisions for manufacturing industry as well as for the agriculture, forestry, and fishery sectors. However, the precise design is rather different from that of, for example, the Danish system, in that the German regulations apply to all companies belonging to the statistical classifications, whereas in Denmark special tax provisions are only granted to specifically defined production processes—and in Denmark industry is still subject to the energy tax only for fuels used for heating purposes. Moreover, in the German scheme industry benefits more than households from the recycling measures, so that the overall tax burden of manufacturing industry has been lowered as a result of the ETR.[7] The beneficial treatment of manufacturing industry was approved by the European Commission in 1999, which ruled that the tax exemptions qualify as state aid for environmental protection and are in accordance with EU law. In particular, they do not contravene the Community guidelines on competition. EU legislation allows the provision of tax exemptions if the sector affected requires 'temporary relief' from environmental taxes.

---

[7] For an ex-ante analysis of the consequences of the German ETR, illustrating that German industry will be a net winner of this reform, see Hillebrand (1999).

38

2.5 The Netherlands

The Dutch government was one of the forerunners in Europe with regard to the introduction of new energy taxes. Up until 2004, four different energy taxes were imposed on the consumption of energy products: the Environmental Tax on Fuels; the Regulatory Tax on Energy, focusing on small-scale consumers; excise taxes; as well as a parafiscal tax, a strategic stockpile fee, known as the COVA levy, on petrol, diesel, gas oil, LPG, and kerosene.

In 1988 the government introduced a general fuel charge which replaced a whole system of programme-specific, earmarked levies in the areas of waste, water, noise, and so on. A further revision took place in 1991 when the general fuel charge was adjusted and became the Environmental Tax on Fuels. The revenues generated by this tax were no longer earmarked and became part of the general budget. The tax still applies and is levied on all energy products used as fuels. This means that energy products used as raw materials and feedstock are exempt from the tax as well as, since 2001, coal and natural gas used for electricity generation. The last revision of this tax took place in 2004, when the tax on energy products, except coal, was incorporated into the system of excise taxes.

The tax base was also subject to several revisions during the 1990s. A CO_2 component was added to the tax base in 1990 and this was revised in 1992, when a new tax scheme was introduced based on the energy and carbon content of the energy products. From 1999, tax rates for all energy taxes have been indexed according to inflation.

In 1996, the Regulatory Energy Tax focusing mainly on small-scale consumers was introduced. This tax is levied on mineral oil products not used for transport purposes, natural gas, and electricity. The purpose of this tax was to stimulate energy efficiency improvements among small energy consumers. The revenues of this tax are recycled back to the economy as part of the Dutch ETR. Competitiveness considerations were given serious thought when the tax was designed. The rationale behind the tax exemption for large industrial energy consumers was the potential risk of harming their export competitiveness with the unilateral introduction of such a tax.

Energy consumed was only taxed up to a ceiling and every consumer received a tax-free allowance (natural gas and electricity). Institutional problems involved in administering tax-free allowances of non-metered mineral oil products led to the removal of the tax-free allowances in 2001.

They were replaced by a fixed tax reduction per electricity connection which amounted to 141 EUR per year in 2001, increasing to 194 EUR in 2005. Furthermore, the specific rule regarding ceilings was abolished on adoption of the EU Energy Taxation Directive in 2004. The energy tax is now levied on energy consumption above the former ceilings, but the rates levied on mineral oil products are around 10 per cent of the standard rates.

ETR in the Netherlands

In 1998, an ETR was implemented based on revenue neutrality. Revenues, generated mainly by the Regulatory Tax on Energy, are recycled back to the economy (households and industry) by applying different recycling measures that took effect in 1999. The recycling measures targeting house-holds were:

- a 0.6 per cent reduction in the income tax rate charged over the first income bracket;
- an increase in the tax-free allowance of 80 HFL (36.3 EUR);
- an increase in the tax-free allowance for senior citizens of 100 HFL (45.4 EUR).

Various recycling options were also in place for Dutch industry:

- a reduction of 0.19 per cent in the wage component paid by employers;
- an increase in the tax-free allowance for small independent businesses (tax credit for self-employed people raised by 1,300 HFL (590 EUR0);
- a reduction of 3 per cent in the corporate tax rate over the first 100,000 HFL (45,378 EUR).

In 2001, revenues raised by the taxes levied on energy products amounted to 3.2 billion EUR, equivalent to approx. 0.7 per cent of GDP.

Special tax provisions for industry

The Dutch energy taxation scheme follows those implemented in other EU member states that grant special tax provisions to industry. However, the Dutch government perceives energy taxes as one of a whole range of policy instruments addressing energy policy issues. Environmental agreements between the government and large-scale energy-consuming industries that commit to improving energy efficiency are of central significance in The Netherlands. These long-term agreements involving energy conservation measures are further strengthened by provisions established in environmental permits.

The specifics of the Dutch energy tax scheme are the tax differentiations with regard to consumption level for natural gas and electricity. These detailed tax rates are unique among EU member states in that tax rates for natural gas are differentiated between seven consumption levels and tax rates for electricity between six levels.

The Dutch ETR addresses both households and industry, which is also reflected in the recycling measures adopted. However, revenues raised by taxes levied on transport fuels are not used within the tax shift programme in The Netherlands as in, for example, Germany.

2.6 Slovenia

The development of energy taxation in Slovenia displays some interesting features, although ETR itself has not been introduced. Nevertheless, it is interesting to look at the situation in this country, as during the last 15 years the system of energy taxation has undergone some fundamental changes. Until 1997, energy products were subject to an *ad valorem* tax ranging from a rate of 5 per cent for natural gas, wood, and district heat, 10 per cent for electricity and coal, and 20 per cent for fuel oil and all other non-transport fuels (a higher rate of 32 per cent applied to high sulphur fuel oil). The tax rates for transport fuels were much higher, amounting to 90 per cent for unleaded petrol, 140 per cent for leaded petrol, and 190 per cent for diesel. The rates of the *ad valorem* taxes had remained constant since the early 1990s and were abandoned in 1997 for almost all energy products, with the exception of taxes levied on transport fuels, which were finally abolished in 1999.

A new system of taxation came into force in 1997 and 1999, respectively, when the Value Added Tax Act and the Excise Duty Act were adopted by the Slovenian parliament. Since this time, excise taxes have been of an *ad quantum* nature. However, this change was accompanied by an increase in the number of taxable energy products; that is, in 1997 only light fuel oil was subject to an excise tax, 1999 then marked the start of an excise tax on transport fuels (petrol and diesel), and in 2000 an excise tax was levied on the use of natural gas. As Slovenia joined the EU in 2004, the EU minimum excise rates established under the Energy Taxation Directive also now apply in Slovenia.

Another feature of the Slovenian energy taxation scheme is interesting, as Slovenia was the first country in Eastern and Central Europe to introduce a CO_2 tax. This tax was implemented in 1997 and applied to all energy products, except coal used for electricity production, which was exempt from the CO_2 tax until the end of 2003. The tax base is a pollution unit and defined in terms of carbon weight, meaning that actual CO_2 tax rates are dependent on the carbon content of the energy products. Initially the tax rate was 1,000 SIT (4.2 EUR) per tonne CO_2 and was raised to 3,000 SIT (12.5 EUR) per tonne CO_2 in 1998. Revenues generated by the CO_2 tax are not hypothecated, although plans were drawn up in 2004 stipulating that around one-third of revenues (around 5 billion SIT) should be used to co-finance investments that promote an increase in energy efficiency and a reduction of CO_2 emissions. Revenues generated from other environmental taxes, such as water consumption tax, waste taxes, and so on, are usually earmarked for specific environmental investment programmes (Máca et al., 2005).

Special tax provisions for business have also been implemented in the Slovenian CO_2 tax regime and companies may be eligible for tax reductions up to 100 per cent. However, the reductions decrease by 8 per cent per annum until the end of the scheme in 2009.

2.7 Sweden

The Swedish energy taxation scheme is very comprehensive and consists of four different types of taxes. Apart from the traditional energy/excise taxes levied on energy products—mainly mineral oil products—in the early 1990s, the Swedish government introduced CO_2 taxes (1991), SO_2 taxes (1991), and a NO_X charge (1992).[8] Since 1995, energy taxes have been indexed for inflation and linked to the consumer price index (CPI).

The scheme displays some rather interesting features and has been amended a number of times over the last 15 years, sometimes as a direct consequence of the fear of harming the competitiveness of Swedish industry. The most striking change was the introduction of the CO_2 tax in 1991. Special tax provisions were not in place for Swedish industry at this time, which resulted in a significant increase in the overall

---

[8] The SO_2 tax and the NO_X charge are not discussed here in detail—see, for further information on these instruments, Nordic Council of Ministers (2002).

carbon/energy tax burden, in particular for energy products other than transport fuels. The introduction of the CO_2 tax was somewhat compensated for by a reduction in the energy taxes. However, a refund mechanism was in place which limited the total energy tax burden paid by industry. The tax scheme was revised in 1993 such that manufacturing industry was completely exempt from energy taxes and paid only a fraction of CO_2 tax rates, reducing the tax burden on manufacturing industry quite dramatically in 1993 compared with 1992. A similar tax switching policy was implemented in 2000 in which energy/excise tax rates were reduced and CO_2 tax rates were correspondingly increased but to a greater degree, leading to an increase in the overall carbon/energy tax burden. It is also worth noting that industry was exempt from paying tax on the consumption of electricity during the period 1993–2003. The decisive factor behind the abolition of this regulation was the EU Energy Taxation Directive, which set minimum tax rates on electricity consumption.

ETR in Sweden

Sweden has embarked on two major fiscal reform processes since the beginning of the 1990s, both involving ETR. The fiscal reform process in 1991 was the first major ETR in Europe.

- The overall objective of the 1991 fiscal reform process was the reduction of personal income taxation by approximately 71 billion SEK (9.5 billion EUR) (approx. 4.6 per cent of GDP in 1991). Income tax rates were cut to around 30 per cent (average rate) and for high-income earners to around 50 per cent. The loss in revenue caused by the reduction in personal income tax was partly compensated for by levying value added tax (VAT) on energy purchases and by introducing the SO_2 and CO_2 tax in 2001. Revenue accrued as a consequence of the new environmental taxes amounted to approximately 18 billion SEK (2.4 billion EUR) (approx. 1.2 per cent of GDP in 1991). This fiscal reform process was not intended to be revenue neutral. However, the ETR component offset some of the shortfall within the national budget.

The second ETR again comprises major fiscal reform and was scheduled for the ten-year period 2001–10. The aims of the Swedish government were set down as follows:

- to lower taxes paid by low- and medium-wage earners, and to encourage adjustment to an ecologically sustainable society (Swedish Government, 2002). As part of this reform process (a so-called green tax shift programme), the intention was to increase revenue generated from environmental taxes by up to 30 billion SEK (3.2 billion EUR) over a ten-year period. This revenue was to be used for budget consolidation in response to the reduction in income tax revenue.

During the first four years of the programme (2001–4), a 10 billion SEK (1.1 billion EUR) tax shift was implemented. For 2005, the planned tax shift involved an increase in environmental tax revenue of around 3.8 billion SEK (410 million EUR), which would only partly offset the shortfall in tax revenue resulting from the planned reductions in labour and capital taxes of approximately 12 billion SEK (1.3 billion EUR). A further increase in environmental tax revenue in the region of 3.6 billion SEK (390 million EUR) was planned for 2006.

Special tax provisions for industry

The Swedish approach of granting special tax provisions to industry passed through different stages. Up until 1992, Swedish industry did not receive any special treatment concerning energy tax rates, that is, industrial energy consumption was subject to the same tax rates as all other economic sectors. However, the total energy tax burden had a ceiling, meaning that the energy tax bill of a company could not exceed 1.7 per cent of sales value. This policy, whereby the difference between the energy tax bill and 1.7 per cent of the sales value was refunded to companies, continued until the end of 1991. In 1992, the ceiling was reduced to 1.2 per cent of sales value and remained in the years to come, but only for the use of coal and natural gas in mineralogical processes. The scheme was finally abolished on 1 January 2007, following the adjustment of tax exemptions for certain industrial processes. A more generous definition of metallurgical processes than that applied earlier was introduced and fuels used in mineralogical processes (i.e. cement, lime, glass) were exempt from tax.

From 1993 onwards, industry was no longer subject to the energy tax but was levied:

- 25 per cent of the CO_2 tax between 1993–7;
- 50 per cent of the CO_2 tax between 1998–2000;

- 35 per cent of the CO_2 tax in 2001;
- 30 per cent of the CO_2 tax in 2002;
- 25 per cent of the CO_2 tax in 2003;
- 21 per cent of the CO_2 tax from 2004 onwards.

Between 2000 and 2005, tax rates generally increased by between 85 per cent (light fuel oil and heavy fuel oil) and around 110 per cent (coal and natural gas). However, for manufacturing industry, the tax increase was only 3 per cent in nominal terms, the large increases being offset by revising the fraction of the CO_2 tax actually to be paid by companies. This fraction dropped from 50 per cent in 1998 to 21 per cent in 2004. This reflects the underlying rationale of the Swedish fiscal reform programme for the period 2001–10 to guarantee and safeguard the competitiveness of Swedish industry.

In addition to being granted generous tax rebates, energy-intensive companies were eligible for a CO_2 tax refund when their CO_2 tax liability exceeded 0.8 per cent of the value of sales, at which point the company was entitled to pay just 12 per cent of the excess tax burden. The refund scheme remained intact from its introduction in 1997 until 1 January 2007. The definition of energy intensiveness has now been adjusted to that in Article 17 of the Energy Taxation Directive. The refund scheme also now contains provisions to ensure payment of the minimum tax levels of the Energy Taxation Directive—this was part of the system already under the Mineral Oils Directive, but later coal and natural gas were added. Currently, companies are required to pay up to 24 per cent of the excess tax burden over the 0.8 per cent limit. It is predicted that around 50 industrial companies are eligible for the refund.

A policy of granting tax-free status to energy products used for electricity generation is in place in Sweden as in other EU member states. However, this policy does not apply to the SO_2 tax, that is, energy products used for electricity production are liable to the SO_2 tax.

From 2004, Sweden has made it possible for energy-intensive industrial enterprises to receive full exemption from tax on electricity if they participate in projects to increase electricity efficiency to achieve the same effect as the tax would have had.

The ambitious ten-year green tax shift programme was stopped by the new centre-right coalition government when it published its budget for 2007. However, the new Swedish government implemented a climate package in the Budget Bill of 2008 which included an increase in the CO_2 tax to 1,010 SEK (108 EUR) per tonne CO_2. The CO_2 tax on fuels used in plants which are covered by the EU Emission Trading Scheme will be reduced in two stages. Up until 2008, all companies were eligible for the special tax provision and paid 21 per cent of the CO_2 tax rate. This provision is increased by 6 percentage points for plants under the EU ETS, meaning that from July 2008 these companies only pay 15 per cent of the nominal CO_2 tax rate. In the second stage, which is planned to start 1 January 2010, the plants are to face a CO_2 tax burden that corresponds to minimum EU tax levels.

2.8 UK

The UK energy tax structure is rather simple when compared to the schemes implemented in the Nordic countries. The scheme relies heavily on the revenues generated from energy/excise taxes levied on transport fuels, in particular. Transport fuel taxes in the UK are among the highest in Europe, and in the world, which can be attributed to the road fuel duty escalator of the 1990s (Ekins and Speck, 2000 and EEA, 2005). A general scheme of energy taxation levied on other energy products does not exist in the UK.

In 1990, the UK government introduced the Fossil Fuel Levy (FFL). The FFL was levied on the purchase of 'leviable electricity' and all consumers faced this levy, that is, the FFL is a tax on electricity. The design of this tax differs because it is an *ad valorem* tax. Initially, the majority of the revenue raised by the FFL was used to subsidize nuclear power and only a smaller fraction was earmarked to support renewables. By the end of 1998, the nuclear industry no longer received subsidies raised by the FFL. Instead, FFL revenues have been utilized to support projects with renewables under the Non-Fossil Fuel Obligation (NFFO). The levy reached its peak in 1992 when the rate was 11 per cent of the end-user electricity price (excl. VAT) and since 2003 the rate has been set to zero per cent; however, this does not mean that the FFL has been abolished.

A new economic instrument was introduced by the UK government in April 2001. This new instrument, the Climate Change Levy (CCL), applies only to non-domestic use of energy (commercial and industrial use). Since 2001, the consumption of natural gas, electricity, and coal is subject to the CCL, and the consumption of LPG is subject to the CCL in addition to existing energy/excise taxes. The revenues generated by the CCL in the UK are used in a programme of tax shifts (ETR).

ETR in the UK

The UK has launched three ETRs within the last decade. In 1996 a landfill tax was introduced and revenues generated from this tax were used to reduce employers' national insurance contributions (NIC). In addition, a small fraction of the revenue was dedicated to a special fund. The fund initially supported investment in waste-related issues as well as research activities in the waste field, but is now largely confined to community projects related to waste. The total tax shift is rather modest and amounted to 0.05 per cent of GDP in the fiscal year 2004/5.

The introduction of the climate change levy (CCL) in April 2001 marked the beginning of the second ETR. The principle of revenue neutrality is adhered to, as the major part of the revenue was used to lower employers' NICs by 0.3 per cent. The remainder is utilized by the Carbon Trust, which was set up to assist investment in energy issues and research activities. The size of the tax shift programme was in the region of 0.06 per cent of GDP (2004/5). The CCL is only levied on natural gas, coal, LPG, and electricity, and the rates remained constant until April 2007, at which point they were increased in line with inflation in the previous year. In 2002, following the first and second tax shift programmes, the UK government introduced the aggregates tax, the revenue from which was used to compensate the reduction in employers' NICs and to establish a special fund ('Sustainability Funds'). This ETR is extremely small in terms of the revenue shift, which is in the region of 0.02 per cent of GDP (2004/5).

Common to all three tax shift programmes implemented in the UK is that they directly target businesses and not the household sector. It is therefore not surprising that in all cases the recycling measure applied reduces the social security contributions which employers have to pay, with the aim of guaranteeing that the total tax burden on business as a whole is unchanged. But this policy clearly leads to differences in tax liability between sectors, that is, some sectors will be net winners and others net losers.

Special tax provisions for industry

Some form of special tax treatment is part of the Climate Change Levy (CCL). Energy-intensive companies are eligible for an 80 per cent tax discount when agreeing to energy efficiency improvement targets. These regulations have been introduced because of concerns that energy-intensive industry in the UK would lose international competitiveness as a consequence of the introduction of the CCL. The approach chosen by the UK government was to give conditional tax exemptions to energy-intensive companies. The concept behind this approach is that companies are entitled to a reduced tax liability when they enter into legally binding Climate Change Agreements (CCA) requiring the adoption of energy-saving measures (OECD, 2005). Of particular interest is the approach selected in the UK of defining energy-intensive industries. It was initially decided that CCA would be limited to those energy-intensive industries which were already registered as energy intensive under the EU Integrated Pollution and Prevention Control (IPPC) Directive (though this stipulation was subsequently relaxed). This clearly limits the special tax provisions to especially energy-intensive industries and contrasts with the German situation, where the process of selecting which industries are eligible for special tax treatment is based on statistical classification, which does not take into account the issue of actual energy intensity. The consequence of this approach is that companies can profit disproportionately from the ETR if they are entitled to tax relief but cannot be described as energy intensive.

2.9 Conclusions

Although the underlying reasons for implementing ETR in EU member states are similar, the designs of the respective tax shift programmes differ. Design varies with regard to the economic sectors affected, as well as the recycling mechanism adopted. However, common to all is that the various reform processes address a twin political objective of environmental improvement (an environmental benefit) and employment support (an economic/employment benefit).

The programmes also exhibit special tax provisions granted to industry because of a fear of losing international competitiveness, which contravenes the recommendations of conventional economic theoretical analysis. However, governments are faced with political constraints and have to make trade-offs between economic efficiency arguments and other aspects, such as distributional outcomes, when making political decisions. Political reality and constraints have made it necessary to grant tax exemptions to manufacturing industries as a prerequisite to implementing ETR in the first place. Nevertheless, this practice can impair achievement of the objectives of ETR by way of excess cost, as the cheapest emission reduction potential need not be exploited. Moreover, beneficial treatment of the industry requires other economic sectors to face higher energy tax rates if a predefined emission reduction goal is to be achieved.

There is no doubt that carbon/energy taxes can have an impact on the competitiveness of energy-intensive industries, although competitiveness is dependent on factors other than just carbon/energy taxes. First, other price factors such as energy import prices and transmission and distribution tariffs (natural gas and electricity), as well as exchange rate variations, have some significance in this discussion. Secondly, non-price factors such as production methods, infrastructure and education are also important. Thirdly and finally, the energy tax burden versus the recycling measure introduced as part of an ETR is a significant factor, and here a detailed analysis of the actual situation is required.

Apart from earmarking some of the additional revenue generated for specific investment programmes to promote and support energy efficiency improvements (Denmark, Germany, and the UK) and reduction of capital taxes (The Netherlands and Denmark), the major part of the revenue is used for the reduction of taxes and charges levied on labour. EU member states made use of the following options:

- reduction in income tax rates; this option is used as a measure to compensate households for higher energy tax bills (Finland, The Netherlands, and Sweden);
- reduction in social security contributions (Denmark, Germany, The Netherlands, and the UK):
  - reduction in employers' social security contributions; this policy reduces labour costs paid by employers. Therefore it is the recycling option which is probably most in line with the political objective of reducing employers' labour costs as a condition for providing new jobs and hence to reduce unemployment;
  - reduction in employees' social security contributions; this policy option affects employees as their net wage increases.

The third and last option is lump-sum transfers to households. The underlying reason for adopting this policy, as implemented for example in The Netherlands, is to compensate that part of society who do not pay income taxes or social security contributions but who face higher energy bills as a result of ETR (e.g. pensioners and students).

The policy decision of which recycling measures actually to implement in EU member states depends crucially on the economic sectors targeted by the ETR. The comparison between the German and UK experience with

ETR illustrates this issue quite clearly. In Germany, all economic sectors, that is, agriculture, industry, trade, public institutions, as well as private households, are faced with higher energy tax bills as a consequence of ETR and all economic sectors have also been compensated in some form. The UK ETR in contrast affects only industrial and commercial energy use; it is therefore not surprising that in the UK only these economic sectors benefit from the recycling measures.

As discussed throughout this chapter, special tax provisions for industries have been implemented in EU member states. However, the specific nature of these provisions varies between countries, making it difficult to provide an overview of the effective tax rates industry experiences. The complex designs of the country-specific as well as industry-specific tax provisions comprise straightforward reduced tax rates for industrial sectors (Denmark, Germany, The Netherlands, Sweden, and the UK), as well as some form of ceiling for the total energy tax burden for individual companies (Denmark, Finland, Germany, and Sweden). Only energy-intensive companies can benefit from the latter, while reduced energy tax rates can be applicable to the whole industrial sector, as in Germany, or to individual industrial sectors and companies—sometimes in combination with agreements requiring these companies to invest in energy efficiency improvements (The Netherlands and the UK). Probably the most striking finding is the fact that the countries with the highest nominal tax rates (Denmark and Sweden) introduced such far-reaching tax provisions that the effective tax rates experienced by industry in these countries are similar to those in the other countries examined. With the exception of taxes levied on electricity consumption, Finnish industry alone faces the same tax rates as other economic sectors, that is, households and the service sector, although the tax rates are lower than in the other countries.

Although there is a long tradition of environmental taxes and charges in environmental policy in Central and Eastern European countries (CEECs), actual ETR, in the sense that an explicitly announced policy shifts the tax burden from conventional taxes, such as labour, to environmentally damaging activities, such as resource use or pollution, has been implemented in just two of the twelve new EU member states, namely Estonia (2005) and the Czech Republic (2008) (Speck, 2007b). However, the countries in the region have revised tax policies and implemented changes in the overall public finance systems during recent years.

These changes have often been linked to the requirements of the EU accession process. Environmental taxes in these countries, in particular those levied on energy products, have moreover been revised during the

1990s (Speck *et al.*, 2001a). Despite the rates of environmental taxes still being relatively lower than those in the six 'old' EU member states which implemented ETR earlier, the revenues achieved by the new member states often reach comparable shares of GDP. However, differences are apparent between the new EU member states, as the environmental-tax-to-GDP ratio has been almost constant in countries such as the Czech Republic and Hungary during the period 1995–2005, whereas in the Baltic countries the ratio more than doubled during the same period, but is still below the average of EU-27 and the newer EU member states. However, as mentioned above, this ratio should only be used with some caution, as a high ratio does not necessarily equate to desirability of the overall fiscal policy in the country in environmental terms.

Environmental tax and charge systems in CEECs require some further attention, as these countries have a long tradition of earmarking revenues from environmental taxes and charges, in particular from pollution and resource taxes, to special parafiscal environmental funds (Speck *et al.*, 2001b). Environmental funds still play a role in co-financing the investments in environmental infrastructure necessitated under the terms of the *environmental acquis*, as the CEECs still face a major backlog with regard to environmental infrastructure implementation.

References

Bach, S. 2004. 'Be- und Entlastungswirkungen der ökologischen Steuerreform nach Produktionsbereichen'. Final report of a research project commissioned by the German Environmental Protection Agency. Deutsches Institut für Wirtschaftsforschung (DIW), Berlin, Germany.

Ekins, P., and Speck S. 2000. 'Proposal of environmental fiscal reforms and the obstacles to their implementations'. *Journal of Environmental Policy and Planning*, 2: 93–114.

European Commission (EC) 2003. Council Directive 2003/96, 'Restructuring the Community Framework for the Taxation of Energy Products and Electricity'. OJ L283/51 (EC).

European Environment Agency (EEA) 2005. *Market Based Instruments in Environmental Policy in Europe*. EEA Technical Report No. 8/2005. Copenhagen, Denmark.

Eurostat. Various years. *Taxation Trends in the European Union*. Luxembourg.

Hillebrand, B. 1999. *Sektorale Wirkungen der Energiesteuerreform, Rheinisch-Westfälisches Institut für Wirtschaftsforschung*. RWI-Papier No. 58. Essen, Germany.

Hoerner, J. A., and Bosquet, B. 2001. *Environmental Tax Reform: The European Experience*. Washington: Center for a Sustainable Economy.

Jensen, A. H. 2001. 'Summary of Danish tax policy 1986–2002'. Working Paper No. 2/2001, Danish Ministry of Finance, Copenhagen, Denmark.

Máca, V., Ščasný, M., and Brůha, J. 2005. 'A prospect on environmental tax reform and energy taxation in new EU member states'. Paper written as part of the COMETR project. Prague, Czech Republic.

Malaska, P., Luukkanen, J., Vehmas, J., and Kaivo-oja, J. 1997. *Environment-Based Energy Taxation in the Nordic Countries*. Ministry of Environment. Helsinki, Finland.

Nordic Council of Ministers. 1994. 'Economic instruments in environmental policy'. *TemaNord*, 1994/647. Copenhagen, Denmark.

—— 2002. 'The use of economic instruments in Nordic environmental policy 1999–2001'. *TemaNord*, 2002/581. Copenhagen, Denmark.

—— 2006. 'The use of economic instruments in Nordic and Baltic environmental policy 2001–2005'. TemaNord 2006:525. Copenhagen, Denmark.

Organisation for Economic Co-operation and Development (OECD) 2005. 'The United Kingdom Climate Change Levy'. Environment Directorate Centre for Tax Policy and Administration. COM/ENV/EPOC/CTPA/CFA(2004)66/FINAL. Paris, France.

Regional Environmental Centre for Central and Eastern Europe (REC) 1999. *Sourcebook on Economic Instruments for Environmental Policy*. Szentendre, Hungary.

Speck, S. 2007a. 'Overview of environmental tax reform in EU member states', in *Competitiveness Effects of Environmental Tax Reforms*. Final Report to the European Commission, DG Research and DG Taxation and Customs Union. National Environmental Research Institute/University of Aarhus, Denmark, 14–78.

—— 2007b. 'Differences in ETR between CEEC and Germany/UK'. Paper presented at the mid-term conference of the Anglo-German Foundation project Productivity and Environmental Tax Reform in Europe (PETRE). November 2007. Berlin, Germany.

—— McNicholas, J., and Markovic, M. 2001a. *Environmental Taxes in an Enlarged Europe*. Szentendre, Hungary: Regional Environmental Centre for Central and Eastern Europe.

———— 2001b. *Environmental Funds in Accession Countries*. Szentendre, Hungary: Regional Environmental Centre for Central and Eastern Europe (REC).

Swedish Government 2002. Revised Budget Statements. PROP. 2001/02:1, Stockholm.

Part II

Industry-Sector Competitiveness

3

Assessing Vulnerability of Selected Sectors under Environmental Tax Reform: The Issue of Pricing Power[1]

John Fitz Gerald,[2] *Mary J. Keeney,*[3] *and Susan Scott*[4]

3.1 Introduction

This chapter describes an analysis of price-setting behaviour by six energy-intensive sectors in six EU countries. The purpose of the analysis is to investigate the relative strengths of world prices and domestic costs in determining the sectors' output prices, with a view to assessing pricing constraints facing the sectors.

The main objective of this study is to assess how a sector would fare under the introduction of carbon taxes or other energy taxes. Such taxes on their own raise domestic costs and the question is to what extent a sector can pass the tax burden on by virtue of its being a price-setter. Alternatively, is the sector a price-taker, meaning that, if it failed to absorb the cost increase, would it be vulnerable to competitive disadvantage under such tax reforms?

To date, in assessing vulnerability to environmental tax reforms and the resulting threats to competitiveness, a number of industrial features

[1] This chapter reproduces an article by the authors with the same title, first published in *Journal of Environmental Planning and Management*, 52/3, April 2009, 413–33. With permission from Taylor and Francis Ltd.
[2] John Fitz Gerald, Head of Economic Analysis Division, Economic and Social Research Institute, Ireland.
[3] Mary J. Keeney, Research Economist, Economic Analysis and Research Department, Central Bank of Ireland, Ireland.
[4] Susan Scott, Associate Research Professor, Economic and Social Research Institute, Ireland.

have been considered, such as energy share, trade exposure, share of the market, market power, and to some extent the potential for improving technological efficiency. Other investigations in this field include Fagerberg (1988), Schroeter (1988), Durand *et al.* (1992), Turner and Van 't dack (1993), Fagerberg (1996), Barker and Köhler (1998), Wolfram (1999), Williams *et al.* (2002), European Commission (2004), ZhongXiang and Baranzini (2004). The purpose of this study is to extend our understanding of 'vulnerability' by considering pricing behaviour, to see how much a sector must find the resources to internally absorb an increase in costs due to environmental taxes.

Concern is expressed that carbon taxes would harm traded energy-intensive sectors by causing their prices to rise out of line with those of competitors in foreign and domestic markets. It is feared that these sectors might cease production or relocate to jurisdictions with lower environmental taxes, or laxer regulations—dubbed pollution havens. Relocation could therefore result in carbon emissions moving elsewhere, with little or no environmental improvement in global terms, merely carbon leakage. A sector with pricing power, however, is not constrained when costs rise and there is less reason to fear that they would cease or relocate.

In the next section, after briefly describing the context for this study, six potentially vulnerable sectors are selected for analysis of their pricing power. The chapter proceeds to summarize the literature on price-setting and formulates a model of price-setting behaviour. The data used and the results of applying the model are then described. After a discussion of results by sector, some implications are outlined, followed by a concluding section. (Appendices are available in the article referenced in footnote 1.)

3.2 Context

The context is the series of environmental tax reforms that were implemented in a number of EU countries, mostly during the period of the 1990s. These tax reforms were the subject of the COMETR project, an *ex post* study of their effects on competitiveness. The reforms in question were the carbon or energy taxes introduced alongside revenue recycling, mainly in the form of reduced labour taxes.[5] Six EU countries introduced such environmental tax reforms (ETRs) (see Table 3.1).

[5] Other modules of the COMETR study have investigated the effects of ETR on greenhouse gas emissions, GDP, and prices, and on uptake of new technology (COMETR, 2007).

Table 3.1. EU countries that introduced carbon/energy tax reforms (ETR countries)

Sweden	1991
Denmark	1995
Netherlands	1996
Finland	1997
Germany	1999
United Kingdom	2001 (announced 1999)

Table 3.2. Potentially vulnerable sectors selected for analysis

	NACE code
Pulp, paper and board	21
Wood and wood products	20
Basic chemicals excl. pharmaceuticals	24 less 24.4
Pharmaceuticals	24.4
Non-metallic mineral products	26
Basic metals	27
Food and beverages	15

Note: Cement forms a large share of non-metallic minerals. The sector food and beverages was included as a comparator.

Given the focus on competitiveness, the sectors deemed potentially most vulnerable and selected for study were those that, in addition to being characterized by high energy intensity, were subject to trade exposure as measured by export and import intensity. Sectors were ranked according to, among other things, energy expenditure as a share of gross value added; the share of exports in the total value of output; and imports as a share of home demand (output plus imports minus exports). Knowledge of specific country characteristics was brought to bear on the selection in order to obtain a balanced representation of sectors, taking into account such issues as the prominence of wood and wood products in the Swedish and Finnish economies. The seven selected sectors were as shown in Table 3.2.

An idea of the vulnerability of these sectors under the introduction of an energy or carbon tax can be gauged by their unit energy costs. Expenditure on energy inputs expressed as a percentage of sectoral gross value added at basic prices is shown in Table 3.3.

It can be seen that, in addition to the expected large variation in energy unit costs between sectors, there is considerable variation across countries at this level of detail. Turning to trade exposure, this is described in

Table 3.3. Unit energy cost in selected sectors in ETR countries, 1998 (% of GVA)

	Wood and paper	Pharma-ceuticals	Basic chemicals	Non-met mineral products	Basic metals	Food, beverages & tobacco	Total gross manuf. output
Denmark	2.4	3.3	4.8	6.8	17.7	5.4	4.6
W. Germany	7.4	19.9	27.2	15.7	56.3	7.8	6.2
Finland	21.4	14.5	19.7	12.3	33.0	4.0	7.9
Netherlands	4.8	24.0	32.3	11.7	29.6	4.7	7.7
Sweden	8.6	15.2	20.4	16.2	29.4	5.8	4.6
UK	4.4	3.8	12.4	8.8	8.5	3.5	4.9
EU15	8.6	17.3	24.4	17.8	42.5	6.8	7.0
ETR (6)	8.0	16.7	24.4	15.0	14.4	6.0	6.5
Non-ETR (6)	9.4	18.9	25.3	21.2	36.3	7.7	7.4

Notes: Annual average exchange rates from Eurostat Ameco database were used. Basic prices are defined as the prices received by producers, minus any taxes payable, plus subsidies received as a consequence of production or sale. The expenditure on energy is made up of the cost in the manufacturing process in each sector of 11 different fuel types: coal, coke, lignite, heavy fuel oil, middle distillates, natural gas, derived gas, electricity, nuclear fuels, crude oil, and steam.
Source: Cambridge Econometrics.

Table 3.4. Proportion of exports to EU destinations, by country (average 1990–1998)

	Wood and paper	Pharma-ceuticals	Basic chemical	Non-met mineral products	Basic metals	Food, beverages & tobacco	Total gross manuf.
Denmark	0.89	0.55	0.54	0.84	0.92	0.70	0.69
W. Germany	0.78	0.63	0.63	0.77	0.69	0.73	0.63
Finland	0.75	0.47	0.46	0.64	0.80	0.39	0.64
Netherlands	1.84	0.67	0.66	0.76	0.76	0.81	0.73
Sweden	0.91	0.81	0.80	0.93	0.87	0.64	0.76
UK	0.61	0.64	0.64	0.56	0.68	0.62	0.64
EU15	0.81	0.66	0.66	0.67	0.75	0.78	0.68

Note: Data recording in the case of pulp and paper for The Netherlands is unreliable.

Tables 3.4 and 3.5 for exports and imports respectively. For reasons that will become clear, it is the share of trade with EU countries shown here that is of special interest.

As shown, the majority of imports were sourced from the EU, and a majority of exports were destined for the EU. At the lower end of trade shares with the EU was the non-metallic mineral products sector (of which cement forms a large share), though, like food, beverages, and tobacco, this sector tended to trade a relatively low share of its output in any event.

Table 3.5. Imports from EU as a share of country imports (average 1990–1998)

	Wood and paper	Pharma- ceuticals	Basic chemical	Non-met mineral products	Basic metals	Food, beverages & tobacco	Total gross manuf.
Denmark	1.02	0.89	0.89	0.95	0.94	0.68	0.77
W. Germany	0.84	0.78	0.78	0.69	0.66	0.79	0.66
Finland	0.88	0.86	0.86	0.68	0.67	0.81	0.70
Netherlands	0.89	0.76	0.76	0.82	0.77	0.77	0.69
Sweden	0.92	0.86	0.86	0.78	0.75	0.83	0.75
UK	0.83	0.83	0.83	0.64	0.68	0.76	0.67
EU15	0.90	0.86	0.86	0.70	0.73	0.81	0.74

3.3 Literature review and price-setting model

Where firms operate in a perfectly competitive market they are price-takers on that market and the price equals the marginal cost of production. If firms' costs are too high, they will just go out of business. However, in many cases, firms may operate under imperfect competition and have a degree of market power. In this latter case, firms may be expected to set their prices as a mark-up on costs (which would include any newly introduced environmental tax), where the extent of the mark-up on cost reflects the demand conditions that they face. Under such market conditions, firms may be able to pass on some of any cost increase (including increased costs arising from environmental taxes) as a higher price. Where firms have market power and are able to discriminate between markets, producers will maximize profits by charging different prices in each market. This is the basis of a measure that is frequently used, the Lerner Index, where the difference between price and marginal cost (as a proportion of price) measures the relative monopoly price distortion, as illustrated for example in Schroeter (1988) and Wolfram (1999).

Price-setting behaviour by firms has been the subject of intensive research in the literature over the past 30 years. Calmfors and Herin (1978) showed that while some Swedish firms exposed to international competition were price-takers, others were less subject to world market prices. Pricing to market is a well-established phenomenon (Krugman, 1987) and there is evidence of its importance in explaining price changes in small open economies (Naug and Nymoen, 1996). Callan and Fitz Gerald (1989) show how Irish firms' pricing decisions changed over the 1980s with the advent of the European Monetary System (EMS) and the growing importance of the EU market; increasingly, Irish firms' pricing decisions were determined by German producer prices (and the bilateral exchange rate). Friberg and Vredin (1997) show how pricing behaviour by Swedish firms evolved over time, with a reduction in the proportion pricing in Swedish crowns and an increase in the proportion invoicing in foreign currencies.

Thus, it is an empirical question, tested in this chapter, whether firms in a particular sector in a particular country are price-takers or whether they have market power, setting their own prices in such a manner that they can pass on at least some of any changes in domestic costs, including taxes.

In this study, the price-setting power of the selected sectors is assessed for the six ETR (Environmental Tax Reform) countries as well as for Ireland. The aim is to understand the global market context and establish, by reference to past behaviour, which sectors can 'pass on' cost increases, such as environmental taxes, and which sectors are constrained to adopt the prices set on world markets.

Two polar cases of the pricing of domestic manufacturing output can thus be posited, where prices are either:

- externally determined, indicating that the sector is a price-taker, or
- determined as a mark-up on domestic costs, revealing that it is a price-setter.

In the latter case, the sector is less exposed to competitive pressures and can be said to have market power. It is less vulnerable in the event of the introduction of the carbon or energy tax, which it can pass on (the revenue recycling side of ETR is left aside). If on the other hand, the former case holds and prices for the sector's product are externally determined, then that sector could indeed be vulnerable in the event of the introduction of a carbon tax, in the absence of adequate mitigating measures such as revenue recycling or if there are no worthwhile technological adaptations that it can undertake. A mixture of the two cases is also a possibility.

In specifying a price-setting model, one may start with a perfectly competitive market, where the law of one price holds. Using p_i to denote the domestic price of sector i's product, and p^f to denote the foreign price expressed in domestic currency, then in the perfectly competitive situation:

$$p_i = p_i^f$$

Meanwhile, in an oligopolistic situation, profit-maximizing firms set prices as an optimal mark-up over marginal costs:

$$p_i = mc_i + \mu_i$$

where mc_i is the marginal cost and μ_i is the mark-up, which can be zero. Leaving aside reactions to short-term events, these relationships should reflect the two sets of influences on the setting of output price. By nesting these two models within a single model, as shown below, we can test whether firms are price-takers or whether they set their price as a mark-up on cost (or whether a combination of these two models is valid):

$$p_i = a_0 + a_1 mc_i + a_2 p_i^f$$

The applicability of the two models to pricing behaviour in individual sectors is tested by checking the statistical significance of the coefficients within this encompassing model. Three outcomes are of interest: the coefficient a_1 on domestic costs is significant, indicating that the firm has market power; a_2 is significant, so that the external price matters and the firm is a price-taker; or they are both significant, indicating that, while the sector has some limited market power, it is heavily constrained by the competitive nature of the global market where it is trading. The equation above is taken to be a long-run price relationship.

It is plausible that, for some sectors, there is room for market power to hold, but there is a limit on the exercise of this power in the long run. This is because, at sufficiently high domestic prices, all markets are contestable such that entry can occur. Obstfeld and Rogoff (2000) show that declining transport costs can have a big impact on relative demand for domestic and foreign goods (thus explaining the falling 'home trade bias puzzle') and hence on relative prices—this could justify changes in pricing behaviour over time.

If estimated coefficients on foreign prices are significant, the sector is likely to be a price-taker and therefore must set its price to match that of its competitors. If the estimated coefficients on only domestic costs are significant, the sector is likely to be much less vulnerable to competition from abroad. Some mixture of the two is possible. Purchasing power parity (PPP) is imposed for the long-run structural relationship between exchange rates and foreign prices.

The basic model to be estimated then becomes:

$$P_d^* = f\left(P_j^f, R_j, W_k\right)$$

where P_d^* is the long-run wholesale price for the sector's domestic output in domestic currency terms; P_j^f is the world wholesale price index in the 'competing' country or bloc j; R_j is the exchange rate with country or bloc j; W_k is the price index for domestic input factor k. Wage rates are used.

The US being a dominant trading bloc, its price is taken as the 'world price' or the price in competing country j. In a second run, the EU price (proxied by the German price) is used as the world or competing country price. To allow for different speeds of adjustment to changes in prices and exchange rates, a lagged response is allowed for, by inclusion of an error-correction type term. The error-correction representation is:

$$\Delta Y_t = a_1 + \lambda (Y_{t-1} - \beta_1 X_{t-1}) + \sum a_2 (i) \Delta y_{t-i} + \sum a_3 (i) \Delta X_{t-i} + \epsilon_{yt} \quad (3.1)$$

where β = parameters of the cointegrating vector, λ is the speed of adjustment parameter where a higher value indicates a faster convergence from short-run dynamics to the long-run situation, and ϵ_{yt} is a white-noise disturbance with no moving average part, and a_i are all parameters.

Equations are estimated for each sector for each country investigated.

3.4 Data

Data are quarterly and run mainly from 1975 to 2002/3, and were sourced from the OECD and Eurostat. There are two basic sources for quarterly data on sectoral output prices, with a sufficient time span. The OECD Statistical Compendium 2004–2, 'Indicators of Activities for Industry and Services ISIC Rev.3' (ceased end 2001) was used to extract producer prices (1995 = 100) for the six countries of interest and for the US price as a proxy for the 'world price'. These prices were available as a domestic price index constructed in national currency. Corresponding domestic producer price indices at the sectoral level were available from Eurostat from 1990 onwards (reference IO7qprin). The OECD series was used after updating with the appropriate rates of change in the price from the corresponding price series up to quarter 4, 2004.

Domestic costs were proxied for each industry by the domestic manufacturing wage in that country. These data are available for the entire period from the OECD and they are calculated as a quarterly index of hourly earnings (2000 = 100) in all manufacturing for each country. Sector-specific wage rates were not available. Owing to the index form of the data, measures such as the Lerner Index are not estimated.

The exchange rates used were obtained from Eurostat (Ameco) and are represented as a quarterly average, where one DM, US dollar, or SEK is expressed in terms of domestic currency units. EUR values post-introduction of the EUR were converted back to domestic currencies existing prior to its introduction in order to achieve a consistent exchange rate time series.

3.5 Results

The basic model in (3.1) above was tested on the data. Table 3.6(a) shows the results and significance levels for the three items, λ (the speed of price adjustment), domestic costs (own country manufacturing wage), and the foreign output price in US dollars. Results are given for the six selected sectors and six ETR countries plus Ireland. A measure of fit is given by the adjusted R^2.

Table 3.6(b) shows the equivalent analysis with the EU (German price) as opposed to the world (US price) to represent the foreign or competing price.

At the base of each table, are two rows headed 'result'. For each sector, these give the number of countries for which the domestic costs and then the foreign price were significant determinants of price.

3.6 Discussion of results by sector

We will be interested to see, first, in which sectors the foreign price is the main influence on price-setting, as this indicates that the sector is a price-taker. By contrast, where domestic costs determine the price, this indicates that the sector has pricing power and, importantly, is thus better able to cope with carbon taxes. Secondly, the question of which foreign price, the world price (proxied by the US price) or the EU price (proxied by the German price), has the most influence is interesting, as it indicates whether the sector competes on the world market, or mainly at the EU level. The foreign price in question in Table 3.6(a) is the US price (as a proxy for the world price). In Table 3.6(b), the foreign price is the German price (as a proxy for the EU price). Even if prices are externally determined, if it is the German price rather than the US price that is significant, this would suggest that an EU-wide application of a harmonized tax would not adversely affect firms' competitive position. That is because the EU price

Table 3.6(a). Modelling the domestic output price—with the US price representing the foreign price [1]

	Chemicals	Food, beverages and tobacco	Non-metallic mineral products	Paper and paper products	Wood and wood products	Basic metals
Denmark						
—Adjustment speed λ	−0.128***	−0.050**	0.009	−0.028	−0.045	−0.062**
—Domestic cost	0.043	0.164	1.377	0.224	0.421	0.174
—Foreign price	0.137	0.295*	−0.920	0.639	0.151	0.643***
—Fit: Adjusted R^2	0.262	0.388	0.540	0.453	0.359	0.323
Germany						
—Adjustment speed λ	−0.137**	−0.012	−0.022	−0.044***	−0.030*	−0.149
—Domestic cost	0.381***	0.242	0.079	0.361***	0.517***	0.270
—Foreign price	0.174	0.517	−0.327	0.244***	0.110	1.246
—Fit: Adjusted R^2	0.492	0.143	0.498	0.732	0.533	0.598
Finland						
—Adjustment speed λ	−0.135**	−0.010	−0.048**	−0.107**	−0.118***	−0.116***
—Domestic cost	0.037	0.745	0.278***	0.285*	0.464***	0.375***
—Foreign price	0.164	0.693	0.056	0.153	0.029	0.301***
—Fit: Adjusted R^2	0.306	0.449	0.410	0.484	0.401	0.600
Ireland						
—Adjustment speed λ	−0.127**	−0.075***	0.041*	−0.087**	−0.150***	−0.400***
—Domestic cost	0.143**	0.340***	0.344*	0.659***	0.572***	0.240***
—Foreign price	0.280**	0.182	−0.013	0.061	0.154**	0.017
—Fit: Adjusted R^2	0.196	0.455	0.394	0.516	0.487	0.213
Netherlands						
—Adjustment speed λ	−0.152***	−0.091**	−0.016	−0.083**	−0.064*	−0.083**
—Foreign price	0.005	0.349***	0.124	0.338***	0.684***	0.300***
—Domestic cost	0.555***	0.123	0.134	0.195*	−0.069	0.405***
—Fit: Adjusted R^2	0.580	0.462	0.395	0.582	0.446	0.508

Sweden						
— Adjustment speed λ	−0.063	−0.017	−0.002	−0.045*	−0.034*	−0.038*
— Domestic cost	0.092	−1.078	−8.456	0.365	0.268	0.410*
— Foreign price	0.590	1.190	0.027	0.604**	0.263	0.711**
— Fit: Adjusted R²	0.246	0.420	0.727	0.612	0.482	0.634
UK						
— Adjustment speed λ	−0.079*	−0.053***	−0.035***	−0.013	−0.067***	−0.055***
— Domestic cost	0.023	0.470***	0.352***	−0.332	0.556***	0.329***
— Foreign price	0.050	0.063	0.260	0.629	0.089	0.267*
— Fit: Adjusted R²	0.195	0.547	0.730	0.742	0.656	0.700
RESULT (no. of significant price determinants in sector)	2 Domestic 2 US	3 Domestic 1 US	3 Domestic 0 US	4 Domestic 3 US	5 Domestic 1 US	5 Domestic 5 US

Notes: [1] Using US$ exchange rates and imposing PPP. * Significant at 10%, ** significant at 5%, *** significant at 1% level.

Table 3.6(b). Modelling the domestic output price—with the German price representing the foreign price [1]

	Chemicals	Food, beverages and tobacco	Non-metallic mineral products	Paper and paper products	Wood and wood products	Basic metals
Denmark						
—Adjustment speed λ	-0.175***	-0.122***	-0.234***	-0.113***	-0.100***	-0.156***
—Domestic cost	0.007	-0.134*	0.513***	0.258***	0.458***	0.079*
—Foreign price	0.389	1.003***	0.139	0.636***	0.358***	0.866***
—Fit: Adjusted R²	0.454	0.429	0.211	0.562	0.420	0.500
Germany	—	—	—	—	—	—
Finland						
—Adjustment speed λ	-0.154***	-0.003	-0.315***	-0.063***	-0.069***	-0.136***
—Domestic cost	0.112	0.327	0.419***	0.197	0.365*	0.194**
—Foreign price	0.210	-6.157	0.053**	0.501	0.186	0.516***
—Fit: Adjusted R²	0.670	0.479	0.227	0.555	0.389	0.643
Ireland						
—Adjustment speed λ	-0.156***	-0.050**	-0.269***	-0.095**	-0.072**	-0.276***
—Domestic cost	0.097	0.327	0.438***	0.429***	0.403***	0.209***
—Foreign price	0.559*	0.069	-0.100	0.500**	0.374*	0.294
—Fit: Adjusted R²	0.172	0.472	0.117	0.580	0.535	0.228
Netherlands						
—Adjustment speed λ	-0.034	-0.098***	-0.177***	-0.031	-0.093*	-0.139***
—Domestic cost	1.610	0.847*	0.406***	0.624	0.703***	0.146**
—Foreign price	-1.874	-1.333	0.412***	0.320	0.065	0.665***
—Fit: Adjusted R²	0.758	0.481	0.178	0.833	0.461	0.605
Sweden						
—Adjustment speed λ	-0.071*	+0.011	-0.176*	-0.079***	-0.029**	-0.124***
—Domestic cost	0.082	3.063	0.716***	-0.013	-0.342	0.047
—Foreign price	1.048*	-1.113	0.018	1.036***	0.806*	0.942***
—Fit: Adjusted R²	0.579	0.553	0.257	0.667	0.587	0.830
UK						
—Adjustment speed λ	-0.113**	-0.056***	-0.167**	-0.020**	-0.049***	-0.115***
—Domestic cost	-0.136	0.306***	0.518***	-0.167	0.324***	0.229***
—Foreign price	0.436*	0.376**	-0.000	0.670**	0.274***	0.476***
—Fit: Adjusted R²	0.540	0.628	0.216	0.774	0.760	0.830
RESULT (no. of significant price determinants in sector)	0 Domestic 3 German	3 Domestic 2 German	6 Domestic 2 German	2 Domestic 4 German	5 Domestic 4 German	5 Domestic 5 German

Note: [1] Using DM exchange rates and imposing PPP. * significant at 10%, ** significant at 5%, *** significant at 1% level would adjust to the higher costs, consequent on environmental taxation, leaving profitability largely unchanged. Under these circumstances, there would be no pressure to move production from its existing location in the EU. However, if the world (US) price dominates, then any environmental tax will tend to put pressure on profitability, increasing pressures to relocate production outside the EU. Results for each of the six sectors are now discussed.

3.6.1 Chemicals

For this sector, there is a better fit generally when the German (EU) price rather than the US (world) price is used to represent the foreign price. The long-run relationship, as measured by λ, was found to be significant for most countries, with a few exceptions.

Turning to the actual strength of domestic versus foreign influences on the output price, results in Table 3.6(a) are somewhat mixed for this sector. The US price is found to influence chemicals output prices only in The Netherlands (quite strongly) and in Ireland. By contrast, in Germany in particular, and in Ireland too, the results suggest that domestic costs have a significant influence, Ireland being influenced by both the US price and domestic costs. In Table 3.6(b), where the German (EU) price was used as the potential foreign price determinant, Sweden and the UK are found to respond to this price, having not responded to the US (world) price. Ireland responds to both foreign prices. Domestic costs are not significant determinants in any country in Table 3.6(b). The speed of adjustment is generally higher where the EU, as opposed to world, price plays the role of external price.

This sector could be vulnerable under an environmental tax regime in certain countries, namely, in The Netherlands and in Ireland, which showed clear signs of taking the US price. The influence of the German price in Sweden, the UK, and also in Ireland suggests that the sector is a price-taker on the 'EU market'. However, if ETR were applied on an EU-wide basis, it would affect EU 'competing' countries in a consistent manner, reducing vulnerability.

3.6.2 Food, beverages, and tobacco

For this sector, the fit is improved when the foreign price is represented by the German, as opposed to the US, price. The adjustment coefficient is also marginally stronger and more significant, though Germany, Finland, and

Sweden are poorly modelled by this long-run relationship, regardless of the foreign price used. Turning to the influences on the domestic output price in Table 3.6(a), only results for Denmark suggest an influence from the US price, though with only 10 per cent significance, while results for Ireland, the UK (quite strongly), and The Netherlands indicate that domestic costs dominate.

In Table 3.6(b), the German price can be viewed as a proxy for the effect of the Common Agricultural Policy on a large share of this sector's prices. We find here that output prices in Denmark and the UK respond to this 'EU price', having not responded to the US price in Table 3.6(a). The UK and The Netherlands show domestic costs exerting a strong influence on their price-setting regimes.

There does not appear to be broad vulnerability to environmental tax reform if applied at EU level therefore. The UK is an example of the third type of outcome mentioned above, where both domestic costs and foreign (German) prices are significant, so that the sector is subject to competitive pressures with respect to European prices, while also responding to domestic cost developments. Were further sectoral disaggregation of data possible, it might clarify this situation, which may arise because of different behaviour in sub-sectors of food processing.

3.6.3 Non-metallic mineral products

This sector is not highly traded and the US (world) price, when used to represent the foreign price, is nowhere significant in explaining movements in the sector's output price. In the UK in particular, the model shows domestic costs as a determinant. If the sector responds to any foreign price, it is likely to respond to the European price. This reflects the low trade shares owing to the bulky nature of the product and its high weight-to-value ratio.

In Table 3.6(b), where the external price is represented by the German (EU) price, the outcome is an inferior fit, however, and the German price is only significant in The Netherlands and to a minor extent in Finland. Domestic costs, on the other hand, significantly determine a substantial portion of this sector's output price in all countries investigated.

To the extent that the external price is at all significant, the fact of it being the German price indicates that a carbon-energy tax applied EU-wide would not create significant competitive disadvantage, given that the rest of the EU would face a similar tax.

3.6.4 *Paper and paper products*

In this sector, we find a better fit when the foreign price is represented by the German (EU) price, rather than by the US (world) price. Nevertheless, Sweden and Germany, and The Netherlands to a minor extent, show a significant impact from the US price, an impact which is large in the case of Sweden according to Table 3.6(a). In Germany's case, domestic costs also have a significant and more dominant impact, a pattern also prevailing in The Netherlands.

Taking the German (EU) price as the foreign price in Table 3.6(b), we find that in size terms and where significant, the external price dominates the influence of domestic cost. This is particularly the case in Sweden, where the relationship with the German price is stronger than with the US price, and also in Denmark and the UK.

This supports the view that this highly traded sector is a price-taker. But, with minor exceptions in Germany and The Netherlands, where the US price is partially influential, the effect on competitiveness would be reduced if ETR applied across the whole of the EU.

3.6.5 *Wood and wood products*

The findings for wood and wood products also show that a better fit is generally obtained using the German (EU) rather than the US (world) price. In all cases that use the German price, the adjustment coefficient is significant, at least at the 5 per cent level. The results for Sweden may be anomalous. For the other countries examined, the coefficient on domestic costs is highly significant and greater in magnitude than that on the foreign currency price.

This suggests a significant degree of market power on the part of firms and an ability to absorb at least some of the incidence of any environmental taxes. The fact that it is the German price rather than that of the US which provides better explanatory power in the equations suggests that, where an environmental tax regime is introduced on an EU-wide basis, there would be little effect on the competitiveness of domestic output. All firms supplying the EU market would be affected in a consistent manner.

3.6.6 *Basic metals*

In the basic metals sector, the US (world) price has a strong and significant influence on output prices, except in the cases of Germany and Ireland.

An even stronger external price effect is found when using the German (EU) price as the foreign price, and this sector is evidently a price-taker on world markets, because results indicate that this sector's pricing is the most responsive to both sets of external prices. Bar the case of Ireland, where neither foreign price has an impact, the German price is a more important determinant of the output price and far outweighs the influence of domestic costs, which in Table 3.6(a) are of lesser significance and in fact insignificant in the case of Sweden. The exceptions, where domestic costs are very significant at the 1 per cent level, are the 'insular' countries, UK and Ireland, though the magnitude of the effect of domestic costs is still smaller than that of the German price.

This indicates that consistent application of environmental tax reform across the EU could temper the effect on competitiveness, though the sector would be vulnerable under a carbon tax nonetheless. The adjustment coefficient suggests a relatively strong and significant, stable long-run pattern of response across all the countries studied.

3.7 Implications

This analysis of price-setting by selected sectors across ETR countries produced plausible results with good explanatory power. Two prices were employed to represent the foreign or competing price, the world price (proxied by the US price) and the EU price (proxied by the German price). Use of the German price generally fitted the data better than the US price. In the case of the non-metallic mineral products sector, it was only the German price that had a significant 'foreign' influence on price-setting. That applied only in The Netherlands and to a very small extent in Finland, suggesting that this sector is at the least vulnerable end of the price-setting spectrum. By contrast, basic metals revealed the most influence from the foreign price and was more likely to be a price-taker and hence vulnerable to domestic cost increases that emanated from environmental tax reform.

Importantly, the results also showed that use of the EU price was in general more consistent with a stable long-run price-setting relationship. Information on trade with the EU, shown in Tables 3.4 and 3.5 above, indicated the predominance of the EU as the source and destination for the products of the selected sectors during the period over which environmental tax reform was being introduced. Therefore the indications are that environmental tax reform introduced on an EU-wide basis (or emissions trading with auctioning) would have a limited effect on the competitiveness of these sectors because all firms supplying the EU market would be affected in a consistent manner.

These time-series regression results can be further employed to rank the selected sectors according to decreasing significance of the external price, that is, in decreasing order of vulnerability or, correspondingly, in increasing order of market power. Thus ranked, the sectors are as follows, starting with the most vulnerable:

- basic metals;
- paper and paper products;
- wood and wood products;
- chemicals;
- food, beverages, and tobacco; and
- non-metallic mineral products.

The basic metals sector was very susceptible to international trading conditions and would be the most affected by an energy or carbon tax. This, of course, is in the absence of mitigating or other measures, such as targeted revenue recycling, technical adaptations, waivers, border tax adjustments, and the like, discussed in COMETR (2007). The sector would face a cost disadvantage compared with its non-EU trading partners (if an EU-wide carbon tax applied) and would not be in a position to mark up its price. At the other extreme, the output price of the non-metallic mineral products sector responded very closely to domestic costs (wage costs in this analysis) and appeared to be relatively insulated from international trading conditions. The study did not show any influence exerted by the world price, proxied by the US price. Of the sectors analysed, non-metallic mineral products would be best placed to absorb a cost increase, such as from carbon or energy taxes, by passing on the tax to its (mostly domestic) customers in the form of higher product prices. Meanwhile, sectors able to make worthwhile alterations to their technology would naturally be better placed still.

While we have established a hierarchy of sectors in terms of their potential vulnerability to environmental tax reform, this hierarchy only holds within a reasonable range of tax rates. It is always possible that in the event of a large rise in tax rates affecting firms' energy prices, firms that were previously price-setters might become price-takers. However, it would take a very sizable rise in tax rates to bring this about.

It is now possible to add the ranking of price-setting power to the criteria used at the outset to gauge a sector's vulnerability under

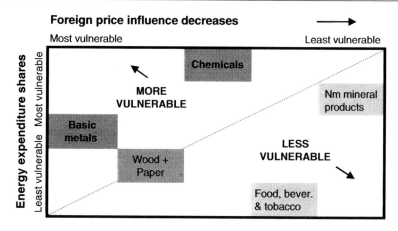

Figure 3.1. Vulnerability with respect to energy expenditure shares and pricing power, ETR countries combined

environmental tax reform. A few examples of combined rankings under various combined criteria are now shown to give a more comprehensive view of the relative vulnerability of sectors. It is noted that the criteria are what the Carbon Trust (2004) terms 'competitiveness drivers' in relation to the EU ETS.

Figure 3.1 illustrates the situation when unit energy costs and pricing power are taken together as two criteria of vulnerability for the combined ETR countries. The vertical axis shows increasing energy expenditure as a share of output, and the horizontal axis shows increasing market power, that is, decreasing foreign price influence in price-setting. Vulnerability is highest in the top left-hand corner, where the energy share is highest and price-setting ability is lowest. Vulnerability is lowest in the bottom right-hand corner. On these criteria, the most vulnerable sectors are basic metals and chemicals in the top left-hand corner of the figure. The chemicals sector has the highest energy expenditure share and basic metals is the most exposed to the world price—it is the least able to pass on cost increases.

In the bottom right-hand corner of the figure are the less vulnerable sectors: food, beverages, and tobacco and non-metallic minerals products. Ranked in the middle in terms of vulnerability is the sector wood and paper.

The implications for policy are that the introduction of ETR would require most care to be paid to its effects on the competitiveness of basic metals and chemicals rather than to non-metallic mineral products, and

Table 3.7. Ranking of sectors with respect to scope for techno-logical adjustment, UK 1995 (with NACE code)

20 + 36	Wood and wood products (least scope, most vulnerable)
27	Basic metals
24	Chemicals
26	Non-metallic mineral products
15	Food and beverages
21	Pulp, paper and paper products (most scope, least vulnerable)

Source: Entec/Cambridge Econometrics (2003).

less again to food, beverages, and tobacco. These rankings of vulnerability apply to the combined six countries that implemented ETR.

As already flagged, another major indication of a sector's vulnerability under carbon taxes is its scope for introducing economically worthwhile energy efficiency investments. Encouragement to use and develop energy efficiency is a prime objective and benefit of carbon taxes, and information on potential technical adjustment was sought as another criterion of vulnerability. Potential technology adjustments that were available to UK energy-intensive sectors had been estimated by Entec, under the Climate Change Agreements procedures and can be used here for illustrative purposes. These adjustment potentials are measured as the sector's percentage energy saving potential at positive net present value. Again, the sectors can be ranked, by scope for adjustment, starting with those that have least scope (i.e. the most vulnerable), as shown in Table 3.7.

The sectors now ranked according to their technological potential for energy efficiency adjustments can be incorporated into a similar figure, Figure 3.2, that relates to the UK. Alongside ranked vulnerability to price competition is shown ranked vulnerability with respect to scope for technological adjustment.

At the extremes, it can be seen that in the UK the basic metals sector is again clearly in a relatively vulnerable position in the figure, now joined by wood and wood products. Food, beverages, and tobacco and the non-metallic mineral products sectors are least vulnerable—they have some modest potential for adapting technology and have some price-setting power. Chemicals and pulp and paper are in between.

These examples give relative placings of sectors and their importance lies in demonstrating that one can rank vulnerability on relevant criteria. They are useful in helping to indicate in which sectors to prioritize mitigation policies to soften any impact on competitiveness in the event of environmental tax reform.

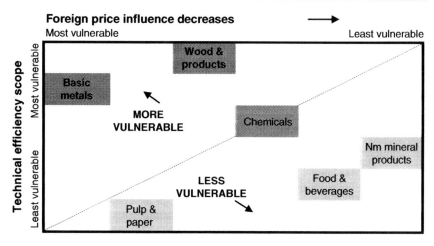

Figure 3.2. Vulnerability with respect to scope for technology adjustments and pricing power, UK

3.8 Summary and conclusions

Six EU member states introduced environmental tax reform (ETR), in the form of carbon taxes with revenue recycling, during the 1990s and after. The purpose of this chapter was to highlight *ex post* the sectors that could be vulnerable under such reform and to explore the nature of their vulnerability. Were they price-takers and, if so, on which markets, and were technological opportunities available that they could call upon in order to reduce vulnerability? Initial screening based on intensities of energy expenditure and other characteristics was undertaken for all sectors to select those six in which price-setting behaviour would be investigated.

A price-setting model was posited and applied in order to throw light on the market power of the selected sectors. The results of the analysis were statistically significant and plausible. The importance of these results is that a sector's price-setting ability, and hence a major aspect of its relative vulnerability, can be realistically assessed.

Among the selected sectors, basic metals had least market power and were most vulnerable, while non-metallic minerals had most power and were least vulnerable. Where the foreign price was a constraint on the price-setting by sectors, it was the EU price (proxied by the German price) that tended to dominate. The importance of this fact is that EU-wide application of environmental tax reform, by contrast with a unilateral application by individual countries, would give less cause for concern about loss of competitiveness. Relocation of production is a feared outcome of the introduction of environmental regulations. An advantage of ETR over environmental regulations lies in the availability of tax revenues that can be used in ways that reduce the inclination to relocate. Any special targeting of revenue recycling and mitigating measures for vulnerable sectors can be refined by using correct criteria, including the market power criterion described here.

The scope for sectors to make profitable adjustments to their technology also has an important bearing on their vulnerability. Energy-saving investment cost curves can be used to assess each sector's scope for adjusting technology, thus enabling them to reduce the adverse effect of the tax side of ETR.

In the analysis, it is the basic metals sector that emerges as being consistently vulnerable on most criteria. This is because it is energy intensive, it is a price-taker on the world market, and its scope for adjusting technology is relatively low. A mitigating factor is its high labour intensity, meaning that any labour tax reduction occurring as part of the ETR could be to its benefit. The chemicals sector shows similar characteristics of vulnerability, though its scope for low cost technology adjustment may be more promising.

The vulnerability of wood and paper depends on the criteria used. In the middle range of vulnerability in terms of pricing power were the wood and wood products and pulp and paper sectors, the former being vulnerable by dint of scarce technology options for improving energy efficiency, while the latter has scope for such adjustments (using evidence from the UK). The non-metallic minerals sector along with food, beverages, and tobacco are the least vulnerable on these criteria of technological potential and pricing power.

References

Barker, T., and Köhler, J. (eds.) 1998. International Competitiveness and Environmental Policies. Cheltenham: Edward Elgar.

Callan, T., and Fitz Gerald, J. 1989. 'Price determination in Ireland: effects of changes in exchange rates and exchange rate regimes'. *Economic and Social Review*, 20/2, Jan.: 165–88.

Calmfors, L., and Herin, J. 1979. 'Domestic and foreign price influences in a disaggregated study of Sweden', in A. Lindbeck (ed.), *Inflation and Employment in Open Economies*. Amsterdam: North-Holland.

Carbon Trust. 2004. *The European Emissions Trading Scheme: Implications for Industrial Competitiveness*. London: Carbon Trust.

COMETR. 2007. 'Competitiveness effects of environmental tax reform'. Policy Brief at <http://www2.dmu.dk/cometr/conference_cont_frame.htm>.

Durand, M., Simon, J., and Webb, C. 1992. 'OECD's indicators of international trade and competitiveness'. Working Papers 120, GD(92)138. OECD: Economics Department.

European Commission. 2004. *European Competitiveness Report*. SEC(2004)1397, Brussels.

Fagerberg, J. 1988. 'International competitiveness'. *Economic Journal*, 98: 355–74.

——1996. 'Technology and competitiveness'. *Oxford Review of Economic Policy*, 12/3: 39–51.

Friberg, R., and Vredin, A. 1997. 'Exchange rate uncertainty and microeconomic benefits from the EMU'. *Swedish Economic Policy Review*, 4/2, autumn: 547–94.

Krugman, P. 1987. 'Pricing-to-market when the exchange rate changes', in S.W. Arndt and J. D. Richardson (eds.), *Real Financial Linkages among Open Economies*. Cambridge, Mass.: MIT Press. 49–70.

Naug, B., and Nymoen, R. 1996. 'Pricing to market in a small open economy. *Scandinavian Journal of Economics*, 98/3: 329–50.

Obstfeld, M., and Rogoff, K. 2000. 'The six major puzzles in international macroeconomics: is there a common cause?' *NBER Macroceconomics Annual*, 15: 339–90.

Schroeter, J. R. 1988. 'Estimating the degree of market power in the beef packing industry'. *Review of Economics & Statistics*, 70/1, Feb.: 158–62.

Turner, P., and Van 't dack, J. 1993. 'Measuring international price and cost competitiveness'. Economic Paper 39. Bank for International Settlements, Basel.

Williams, E., Macdonald, K., and Kind, V. 2002. 'Unravelling the competitiveness debate'. *European Environment*, 12: 284–90.

Wolfram, C. D. 1999. 'Measuring duopoly power in the British electricity spot market'. *American Economic Review*, 89/4: 805–26.

ZhongXiang, Z., and Baranzini, A. 2004. 'What do we know about carbon taxes? An inquiry into their impacts on competitiveness and distribution of income'. *Energy Policy*, 32: 507–18.

4

Trends in the Competitiveness of Selected Industrial Sectors in ETR Countries

Roger Salmons[1] and Alexandra Miltner[2]

4.1 Introduction

By definition, if an environmental tax reform (ETR) in a country raises the production costs of its constituent firms relative to those of competitors in other countries, then it has a negative impact on competitiveness compared to the hypothetical situation where the ETR had not been implemented—all else being equal.[3] However, from a policy perspective, it is often more relevant to consider whether there is a deterioration in competitiveness compared to the actual situation before the ETR was implemented. This will depend on the magnitude of the 'ETR impact' relative to the impacts of other factors that affect competitiveness. For example, if the prices of raw materials or components are falling, or if production efficiency is improving, then unit production costs may fall despite the negative impact of the ETR. Even if the ETR does cause unit production costs to increase, there may be an improvement in competitiveness if the costs of producers in other countries are rising more quickly

[1] Dr Roger Salmons, Visiting Research Fellow, Environment Group, Policy Studies Institute, United Kingdom.
[2] Alexandra Miltner, Research Fellow, Environment Group, Policy Studies Institute, United Kingdom.
[3] If the ETR causes technological (or managerial) improvements in efficiency that would not otherwise have occurred, then the ETR may cause unit production costs to fall and hence competitiveness to improve—an extension of the so-called 'Porter Hypothesis'.

(for whatever reason). Conversely, there may be a deterioration in the competitiveness of a sector even if the ETR impact is positive.

In recognition of this, the objective of this chapter is to assess whether there is any indication that the seven European Union member states which implemented energy-related tax reforms between 1990 and 2001 suffered any losses of sectoral competitiveness over the period.[4] The seven countries are Germany (DE), Denmark (DK), Finland (FI), the Netherlands (NL), Sweden (SE), Slovenia (SI), and the United Kingdom (UK). Details of the tax changes that were introduced under the ETRs in each country are provided in Chapter 2. The following eight industrial sectors—defined at the NACE-3 level of aggregation—were selected for the assessment, and for the detailed econometric analysis in the next chapter.[5] These were chosen specifically to span a range of different energy and trade intensities.

- Food and food products (NACE 15.1)
- Paper and paper products (NACE 21.2)
- Basic chemicals (NACE 24.1)
- Pharmaceuticals (NACE 24.4.)
- Glass and glass products (NACE 26.1)
- Cement, lime and plaster (NACE 26.5)
- Ferrous metals (NACE 27.1–3)
- Non-ferrous metals (NACE 27.4)

The chapter is divided into two distinct, but related, parts. In Section 4.2, a number of potential competitiveness indicators are assessed in the context of a formal theoretical model to determine how well they reflect actual changes in competitiveness. In Section 4.3, those indicators that have been shown to be valid are used in an empirical assessment to determine whether any of the seven ETR countries suffered a deterioration in competitiveness across the selected sectors over the period 1990–2002.

Underlying the empirical assessment is a series of country data sets that were collated during the COMETR project. Each data set contains generic data for the country, such as exchange rates and emission factors, and sector-specific data for the selected sectors on energy use, energy prices and taxes, economic variables, and labour market variables. Wherever possible, data were collated directly from official sources (i.e. national

---

[4] That is, where taxes were introduced, or increased, on energy products or carbon emissions, with the revenues being used to reduce employer or employee labour taxes.

[5] The ferrous metals sector is defined as the aggregation of three NACE-3 sectors: 27.1, 27.2, and 27.3.

statistical agencies, Eurostat, IEA, etc.), either from published data sets or through bespoke analyses commissioned from national agencies. However, in some cases it was necessary to combine data from different sources; to adjust data where there were obvious errors or inconsistencies; or to estimate missing data for certain years. The resultant data sets are believed to provide the most comprehensive and consistent economic and energy-related data for the selected sectors, although as one would expect at this level of aggregation (i.e. NACE 3), individual time series are subject to a high degree of volatility from year to year.

4.2 Theoretical assessment of competitiveness indicators

A firm suffers a loss of price competitiveness if its unit cost of production rises faster than its competitors', or if it falls more slowly. Similarly, at the aggregate level, a sector in one country becomes less competitive if the production costs of its constituent firms rise faster (on average) than those in other countries, or fall more slowly. In theory, therefore, changes in sectoral competitiveness could be measured directly by comparing the changes in unit production costs of firms in all competing countries. Unfortunately, in practice, there are a number of problems with this direct approach.

In many sectors, there may be no meaningful unit of measurement for aggregate output (even at the firm level), making the definition of unit production cost problematic. While it is possible to use economic output indicators—such as gross output or gross value added (in constant prices) as a proxy for physical output—the strength of the correlation between the measures is open to question.[6] Even for sectors where there is a meaningful unit of measurement for output (e.g. tonnes of cement), data may not be publicly available for all—or indeed any—of the competing countries. Furthermore, where data do exist, they will usually only be available at an aggregate level, not for individual firms.

Consequently, the direct assessment of changes in competitiveness is likely to be either unreliable (at best), or impossible (at worst). What is needed therefore is a 'proxy variable' (or set of variables) whose movements are closely correlated with changes in competitiveness, but that does not suffer from these practical measurement problems. Intuitively,

---

[6] See Freeman *et al.* (1997); Worrell *et al.* (1997); Bernard and Cote (2002) for a comparison of physical and economic output indicators in the context of measuring sectoral energy intensity.

one might expect that if a sector's competitiveness improves, then its *share of global production, export intensity,* and *profitability* (as a percentage of sales) would all increase, while its *import penetration* would fall; with the opposite being the case if it deteriorates. If this expectation is valid, then any, or all, of these readily available measures would provide a reliable indicator of changes in competitiveness. However, before they can be used, it is necessary to determine whether the intuition is indeed true. In order to do this, each of the four indicators is assessed within the framework of a formal theoretical model.

4.2.1 Model definition

The model used to assess the validity of the potential indicators is a generalization of the 'reciprocal dumping' model used in international trade analysis (see Brander, 1981; Brander and Krugman, 1983). A homogeneous product is produced by individual *firms* $i \in I$, located in *countries* $j \in J$; with the subset of firms located in a particular country being denoted by I^j and the number being denoted by the parameter N^j. The product is purchased by consumers located in distinct *markets* $k \in K$; where the latter are defined geographically. For simplicity, it is assumed that there is a one-to-one correspondence between countries and markets.[7] Consequently, the number of markets is equal to the number of countries; with producers in each country having a 'home market' and $K - 1$ 'export markets'. In each market, the firms compete as quantity-setting Cournot oligopolists facing a market-specific linear inverse demand function: $P^k = a^k - b^k X^{\cdot k}$, where P^k is the market price, $X^{\cdot k}$ is the aggregate value of market sales, and a^k is the 'choke price' (i.e. the price at which sales fall to zero).[8]

All firms have a constant unit cost of production $\left(c_i^j\right)$ and a fixed cost $\left(F_i^j\right)$, both of which differ between firms. In addition, they incur market-specific constant unit transportation costs (r^{jk}) which are assumed to be the same for all firms in a particular country. For simplicity, it is assumed that the transportation costs are equal to zero for all home-market sales.

---

[7] This does not have to be the case, so long as each market is distinct. For example, there may be several regional or local markets within a country, or a single market may span several countries. However, this complicates the definitions of countries' import penetration and export intensity.

[8] With distinct (i.e. segmented) markets and Cournot conjectures, firms make separate quantity choices for each market under the assumption that all other firms hold their output constant.

The following three definitions of average production cost are used in the analysis:

- average of unit production costs for firms in country $j \in J$

$$c^j = \frac{1}{N^j} \left(\sum_{i \in I^j} c_i^j \right) \tag{1a}$$

- average production cost of sales in market $k \in K$ by firms located in country $j \in J$

$$\check{c}^{jk} = \frac{1}{X^{jk}} \left(\sum_{i \in I^j} c_i^j x_i^{jk} \right) \tag{1b}$$

- average of unit production costs for all firms

$$\hat{c} = \frac{1}{N} \left(\sum_{j \in J} \sum_{i \in I^j} c_i^j \right) = \frac{1}{N} \left(\sum_{j \in J} N^j c^j \right) \tag{1c}$$

where X^{jk} and x_i^{jk} are respectively the aggregate and individual sales in market $k \in K$ by firms located in country $j \in J$. Under the assumption of country-specific unit transportation costs, within each country, firms with lower unit production costs will have higher market shares in all markets and hence it follows directly that $c^j > \check{c}^{jk}$.

The competitiveness of a firm is defined to improve if the increase in its unit production cost is less than the average increase in unit production costs of all other firms, or if the reduction in its unit production cost is greater than average reduction of the other firms. Conversely, it will lose competitiveness if its unit production cost increases by more than its competitors, or reduces by less. This definition can be extended to the sectoral level, with a country gaining competitiveness if the average increase (decrease) in the unit production costs of its constituent firms is less (greater) than the average increase (reduction) of firms in all other countries, and losing competitiveness if the reverse is true.[9] Formally, the necessary and sufficient condition for sector $j \in J$ to gain (lose)

---

[9] By definition, the average change in unit production costs of constituent firms (i.e. $\Sigma(\Delta c_i^j)/N^j$) is the same as the change in the average of the unit production costs (i.e. $\Sigma(\Delta c_i^j/N^j)$).

competitiveness is that:

$$\frac{1}{N^j}\left(\sum_{i\in I^j}\Delta c_i^j\right) \quad <(>) \quad \frac{1}{(N-N^j)}\left(\sum_{i\notin I^j}\Delta c_i^j\right)$$

which can be rearranged to yield the following condition:

$$\Delta c^j - \Delta\hat{c} \quad <(>) \quad 0 \tag{C1a}$$

For small changes in unit costs, the condition can be expressed in terms of differentials:

$$dc^j - d\hat{c} \quad <(>) \quad 0 \tag{C1b}$$

4.2.2 Assessment of potential competitiveness indicators

The validity of each of the four measures as an indicator of changes in competitiveness for a particular country is assessed by deriving a necessary and sufficient condition for the measure to increase, expressed in terms of the average change in unit production costs for that country relative to the global change (i.e. $dc^j - d\hat{c}$).[10] If this condition coincides with the competitiveness condition (C1b), then the measure is a robust indicator of changes in competitiveness. However, if it diverges significantly, then it is a poor indicator that should be used with caution, or not used at all.

In addition to changes in unit production costs, the analysis allows for the possibility that markets may be growing or shrinking over time. This is achieved by allowing the choke price in each market (a^k) to vary independently, while keeping the slope of the inverse demand curve (b^k) constant.[11] All of the other exogenous parameters in the model are held constant. In particular, the number of firms in each sector is fixed—that is, there is no entry or exit as a result of the changes in unit costs.

Share of global production

A necessary and sufficient condition for share of global production (σ^j) of country $j \in J$ to increase (decrease) is that:

$$dc^j - d\hat{c} \quad <(>) \quad \frac{1}{N+1}\left(1 - \frac{\sigma^j}{\rho^j}\right)(d\hat{a} - d\hat{c}) \tag{C2}$$

---

[10] Details of the derivation of the conditions are provided in the Final Project Report, which can be downloaded from:<http://www2.dmu.dk/cometr/COMETR_Final_Report.pdf>.

[11] Thus, the (inverse) demand curve may shift outwards (or inwards) over time, but the slope does not change.

where ρ^j is the country's share of total number of firms (i.e. N^j/N) and

$$d\hat{a} = \left(\sum_{k \in K} \frac{d_{ak}}{b^k}\right) \bigg/ \left(\sum_{k \in K} \frac{1}{b^k}\right)$$

The expression on the right-hand side of condition (C2) is not generally equal to zero—only being so if the country's share of global production is equal to its share of the total number of firms (i.e. $\sigma^j = \rho^j$); or if the weighted average change in the choke prices in the various markets is equal to the average change in the unit costs of all firms.[12] Neither of these two conditions is likely to be true in general, although the country shares of global production and number of firms will be similar if there is little variation in the average scale of firms between countries. In general, the expression may be positive or negative. However, if the total number of firms is relatively large, then the right-hand side of (C2) will be approximately equal to zero.[13] For most sectors, this is likely to be the case and hence the direction of change of a country's share of global production provides a good indicator of its change in competitiveness. If a sector gains competitiveness, its share of global production increases; if it loses competitiveness, it declines (see Figure 4.1[14]).

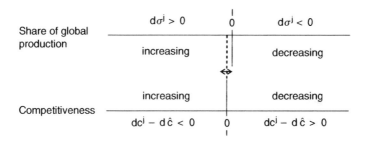

Figure 4.1. Changes in competitiveness and global market share

---

[12] The changes in the choke prices are weighted by the slopes of the respective demand curves (i.e. $1/b^k$).

[13] The assumptions that are made in the derivation of (C2) impose an upper bound on the global number of firms (N), which is dependent on the relative values of the choke price and the unit costs (including transport costs) in each market. Details are provided in the Final Project Report.

[14] In Figures 4.1–4.3, it is assumed that the value of the right-hand side of the condition is positive. However, this need not necessarily be the case.

Import intensity of home market

A necessary and sufficient condition for import penetration of the home market (μ^j) for country $j \in J$ to decrease (increase) is that:

$$dc^j - d\hat{c} \quad < (>) \quad \frac{1}{N+1}\left(1 - \frac{(1 - \mu^j)}{\rho^j}\right)(da^j - d\hat{c}) \qquad (C3)$$

The right-hand side of condition (C3) is only equal to zero if the home market share of the country's producers $(1 - \mu^j)$ is equal to the country's share of the total number of firms, or if the change in the choke price in the home market is equal to the average change in the unit costs of all firms. Unlike the previous indicator, one might expect a country's share of its own market to be significantly greater than its share of total firms (implying that the middle term is negative). Furthermore, unless the change in the choke price is the same for all markets, the magnitude of the final term in (C3) will—by definition—be greater than the corresponding term in (C2) for some of the countries (but smaller for others). Consequently, the absolute magnitude of the right-hand side of (C3) is likely to be larger than for (C2)—at least for some countries. However, again if the total number of firms is relatively large, then it will be approximately equal to zero, and hence the direction of change of the import intensity of a country's home market is also likely to provide a good indicator of its change in competitiveness. If a sector gains competitiveness, the import intensity of its home market decreases; if it loses competitiveness, it increases (see Figure 4.2).

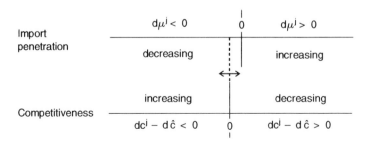

Figure 4.2. Changes in competitiveness and import penetration of home market

Export intensity

A necessary and sufficient condition for the export intensity (ξ^j) of country $j \in J$ to increase (decrease) is that:

$$dc^j - d\hat{c} \begin{cases} < (>) & \left(\frac{1}{N+1}\right)(d\tilde{a}^j - d\hat{c}) \quad \text{if} \quad \xi^j < \eta^j & \text{(C4a)} \\ > (<) & \left(\frac{1}{N+1}\right)(d\tilde{a}^j - d\hat{c}) \quad \text{if} \quad \xi^j > \eta^j & \text{(C4b)} \end{cases}$$

$$\text{where} \quad d\tilde{a}^j = [(1-\xi^j)d\hat{a} - (1 - \eta^j)da^j]/(\eta^j - \xi^j)$$

$$\eta^j = 1 - \frac{1}{K}\left(\frac{1}{b^j} \Big/ \left(\sum_{k \in K}\frac{1}{b^k}\Big/K\right)\right)$$

Again, if the total number of firms is large, the value of the expression on the right-hand side of the inequality is approximately equal to zero. However, as can be seen in Figure 4.3, the relationship between changes in export intensity and changes in competitiveness is not as straightforward as for the previous two indicators.

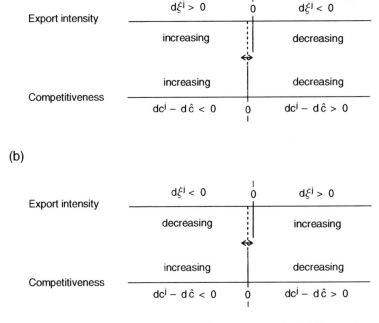

Figure 4.3. Changes in competitiveness and export intensity (a) Export intensity less than threshold value (η^j) (b) Export intensity greater than threshold value (η^j)

If the country's export intensity is less than the parameter value η^j, then it increases as it gains competitiveness and declines as it becomes less competitive. However, if it is greater than the parameter value, then the opposite is the case. The value of η^j is determined by the slope of the demand curve in the country's home market relative to its export markets and the total number of markets. If there is a large number of markets with similar slopes for their respective demand curves, then the value of η^j will be close to one; in which case export penetration increases as a country gains competitiveness.[15] However, if there are only a few markets with significant variation in the slopes of demand curves, a country with high export intensity may experience a decline as its competitiveness increases. Consequently, care should be taken to determine which case applies when using movements in export intensity as an indicator of changes in competitiveness.

Profitability as a percentage of sales

The overall profitability of country $j \in J$ is equal to the weighted average of its operating profitability in each market, less aggregate fixed costs divided by its total revenue; where the weights are equal to the markets' respective shares of total country revenue. Consequently, the change in overall profitability can be decomposed into three parts, due respectively to:

- the change in the country's profitability in each of the individual markets;
- the change in the mix of the country's revenues across the various markets;
- the change in the significance of its fixed costs.

A necessary and sufficient condition for the market profitability (π^{jk}) of country $j \in J$ in market $k \in K$ to increase (decrease) is that:

$$dc^j - d\hat{c} \; < (>) \quad \left(\frac{1}{N+1}\right)(da^k - d\hat{c}) - \Lambda^{jk} \tag{C5a}$$

$$\text{where} \quad \Lambda^{jk} = \left(\frac{\pi^{jk}}{1 - \pi^{jk}}\right)dc^j + \left(\frac{1}{1 - \pi^{jk}}\right)(c^j - \check{c}^{jk})\frac{dX^{jk}}{X^{jk}}$$

$$dX^{jk} = \frac{N^j}{b^k}\left(\left(\frac{1}{N+1}\right)(da^k - dc^j) - (dc^j - d\hat{c})\right)$$

---

[15] This assumes that there is no re-export of imports and hence—by definition—export intensity must be less than one.

Unlike the previous indicators, the right-hand side of condition (C5a) does not necessarily tend to zero as the number of firms increases. Depending on the value of Λ^{jk}, it may be significantly positive or significantly negative. Consequently, it is possible that a country may suffer a fall in profitability in some (or all) of its markets when it gains competitiveness. Conversely, profitability may increase despite the country losing competitiveness.

If unit production costs within a country are homogeneous (i.e. $c^j = \check{c}^{jk}$) and increasing, then the right-hand side of (C5a) will generally be negative, as the positive value of Λ^{jk} is likely to dominate the first term if the total number of firms is large. Hence market profitability will only increase if the country is gaining competitiveness (see Figure 4.4). Similarly, if unit production costs are falling, then the right-hand side will generally be positive and hence market profitability will only fall if the country is losing competitiveness. However, these are only sufficient conditions—one

(a)

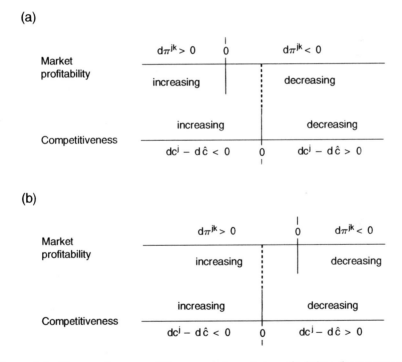

(b)

Figure 4.4. Changes in competitiveness and market profitability (homogeneous production costs) (a) Increasing average unit production cost (b) Decreasing average unit production cost

87

cannot infer anything about the change in competitiveness if profitability is falling in the first case, or rising in the second. Furthermore, homogeneity of unit production costs is a very strong assumption that is unlikely to be justified in practice.

A change in the mix of the country's revenues between markets will have no impact on overall profitability if there is no variation in profitability between markets, or if all markets experience the same growth in revenue. However, in general, this will not be the case and the overall sign of the mix effect will depend on the relative percentage changes in revenue for each market. If revenue growth is greatest in markets with below average profitability, then the overall impact will be negative. If the reverse is true, it will be positive. The relative growth rates will depend on a range of factors (i.e. parameter values) and it does not follow that an increase in competitiveness will necessarily lead to greater (or smaller) increases in revenue in more profitable markets. Consequently, there is no predictable relationship between changes in competitiveness and changes in profitability due to mix effects.

Finally, the change in profitability due to the change in the significance of the aggregate fixed cost depends on the change in the country's total revenue. In the special case where there are no changes to the values of the choke prices in any of the markets (i.e. $da^k = 0$ for all $k \in K$), then if average unit costs are rising globally (i.e. $d\hat{c} > 0$), a sufficient condition for the fixed cost effect to be positive (i.e. improve overall profitability) for country $j \in J$ is that:

$$dc^j - d\hat{c} < \quad - \left(\frac{1}{N+1} \right) d\hat{c} \tag{C5b}$$

Similarly, if unit costs are falling globally (i.e. $d\hat{c} < 0$), then a sufficient condition for the fixed cost effect to be negative for country $j \in J$ is that:

$$dc^j - d\hat{c} > \quad - \left(\frac{1}{N+1} \right) d\hat{c} \tag{C5c}$$

However, while these conditions are sufficient, they are not necessary. Furthermore, they rely on the assumption that the markets are static (i.e. the choke prices do not change in any markets). In general, the overall impact on a country's total revenues is unclear. They may fall in some (or all) of its markets when it is gaining competitiveness, and they may rise when it is losing competitiveness. So again there is no predictable relationship between changes in competitiveness and changes in profitability due to fixed cost effects.

Thus, in general all three components of the change in profitability can move in either direction as a sector gains or loses competitiveness. Only under a number of restrictive assumptions can changes in a country's profitability provide a good indicator of changes in competitiveness.

4.3 Empirical assessment of competitiveness trends

The theoretical analysis in the previous section suggests that changes in global market share and changes in import penetration are both likely to provide good indicators of changes in a country's competitiveness. Changes in export intensity can also provide a good indicator, provided that one knows whether export intensity is above or below a threshold value. Consequently, these three indicators are used to assess whether there is any evidence that competitiveness deteriorated over the period 1990–2002 for any of the seven ETR countries in the selected industrial sectors. In order to do this, the 'theoretical indicators' from the model have to be translated into 'practical indicators' for which empirical values can be constructed using the economic and trade date collated in the COMETR data sets for each country. While this is relatively straightforward, there are a number of issues that need to be addressed.

According to the theoretical analysis, one should expect import penetration to rise and global market share to fall if a country loses competitiveness; with the opposite being the case if it gains competitiveness. Calculation of import penetration values is straightforward, these being equal to the ratio of a country's total imports of goods and services to its domestic demand (which is equal to total output less net exports). However, the calculation of global market share values is more problematic. The first problem is that the sectors produce a heterogeneous range of products and while it is possible to get global production share information for individual products in some of the sectors, these do not necessarily provide a good picture of the aggregate changes in market share. An obvious answer is to calculate aggregate production share values based on the economic value of gross output, valued at constant prices. On this basis, a necessary and sufficient condition for a country to increase its share of production in a particular sector is that its growth in real output is greater than the weighted average growth rate for all countries against which it competes. Consequently, changes in production share can be assessed by comparing changes in the ETR countries' respective indexes of production (the values of which are collated in COMETR data sets) with changes in the global index of production.

This leads to the second problem—which countries should be included in the global production index? Ideally, one would want to include all those countries against which the ETR countries compete in any market in the world. Unfortunately, while it is possible to identify these from international trade data, production index values are not available for countries outside the European Union at the NACE 3 level of aggregation needed for the assessment. Consequently, the EU25 production index values (taken from Eurostat) are used as a proxy. For most sectors, intra-EU25 imports and exports account for around 70–85 per cent of the respective total values across the ETR countries and hence the use of this proxy may not be unreasonable. However, for basic chemicals, pharmaceuticals (exports), and aluminium (imports), the rest of the world accounts for a larger share of total trade (i.e. around 30–50%) and hence caution should be exercised in interpreting the indicator for these sectors.

Export intensity values are calculated directly from the collated data as the ratio of a country's exports to its gross output. However, as was discussed in the previous section, the relationship between changes in competitiveness and changes in export intensity depend on whether export intensity is less than, or greater than, a threshold value (η_j) which is determined by the slope of the country's home demand curve relative to its export market and the number of markets in which it competes (see condition C4 above). Hence, in order to interpret any trends in export intensity, it is necessary to make some assumptions about these underlying parameters. If one assumes that—as a minimum—every country in the EU25 competes in every other member state, and that the slope of any country's demand curve is no greater than double the average across all of the member states (i.e. the sensitivity of demand to price changes is not too heterogeneous), then a lower bound for the threshold value is 0.92 (i.e. $1 - 2/25$). For the vast majority of cases considered, export intensity is considerably below this value and hence increases in export intensity can be interpreted as indicating an improvement in competitiveness. There are a couple of instances where export intensity is well above 100 per cent; indicating either that imports are being re-exported, or that there are inconsistencies between the production and trade data sources. Since the model does not allow for re-exports, the conclusions from the theoretical analysis may not be valid. Consequently, export intensity is not used as an indicator of changes in competitiveness in these cases.

For each of the 56 cases (i.e. seven countries for each of the eight sectors), trend lines for the three indicators over the period 1990–2002 are fitted by linear regression. In some cases, lack of data means that it is not possible to calculate an indicator value. In particular, production index values are not available for Slovenia for any of the eight sectors, while import data are not available for Finland for five of the sectors, and hence it is not possible to construct the corresponding indicators in these cases. The signs of the resultant trend line slopes are shown in Tables 4.1–4.8 for those that are statistically significant at the 10 per cent level; a positive sign indicating an upward trend and a negative sign indicating a downward trend. For those cases where the slope is significant, the cell is shaded dark grey if it indicates that there was a gain in competitiveness, and light grey if it indicates that there was a loss.

Ideally, all three indicators should show a consistent picture. For example, a negative slope for import penetration should be accompanied by a positive slope for relative production and for export intensity. However, in practice, this is rarely the case. This may be due to the poor quality of some of the data at this level of aggregation; to the compromises that

Table 4.1. Summary of indicator trends—meat and meat products

Country	Import penetration	Relative production	Export intensity	Overall
DE	−		+	Gain
DK	+	−	+	Loss
FI	n/a	+		
NL	+		+	
SE	+	+		
SI		n/a		
UK	+		−	Loss

Table 4.2. Summary of indicator trends—paper and paper products

Country	Import penetration	Relative production	Export intensity	Overall
DE	+	−	+	Loss
DK	−	−		
FI	n/a	+	−	
NL	+	−	+	Loss
SE	−	+	−	Gain
SI	+	n/a	+	
UK	+	−	+	Loss

Table 4.3. Summary of indicator trends—basic chemicals

Country	Import penetration	Relative production	Export intensity	Overall
DE	+	–		Loss
DK		...		
FI	n/a	+	+	Gain
NL		–		
SE		...		
SI		n/a		
UK		–		

Table 4.4. Summary of indicator trends—pharmaceuticals

Country	Import penetration	Relative production	Export intensity	Overall
DE	+	–	+	Loss
DK	...	+	–	Gain
FI	+	–	+	Loss
NL	+		+	
SE		+		
SI	+	n/a	+	
UK	+		+	

Table 4.5. Summary of indicator trends—glass and glass products

Country	Import penetration	Relative production	Export intensity	Overall
DE	+	–	+	Loss
DK	+	+		
FI	n/a	+		
NL	n/a	+	n/a	
SE	...		–	Loss
SI	...	n/a	–	
UK		–		

have been made in translating the theoretical indicators into empirical measures; or to the fact that they are not perfect indicators of competitiveness even in theory. Whatever the reason, it means that a decision rule is needed in order to draw any conclusions about changes in competitiveness. The rule used for this assessment is that at least two of the indicators should show a consistent picture, and the conclusions from applying this rule are shown in the final column of each table. This is a relatively

Table 4.6. Summary of indicator trends—cement, lime, and plaster

Country	Import penetration	Relative production	Export intensity	Overall
DE	−	−	+	Gain
DK	−	+	−	Gain
FI	n/a		n/a	
NL	n/a	+	n/a	
SE				
SI	+	n/a		
UK	+	−		Loss

Table 4.7. Summary of indicator trends—ferrous metals

Country	Import penetration	Relative production	Export intensity	Overall
DE	+	+	+	Gain
DK				
FI		+		
NL		+		
SE	+	+	+	Gain
SI	+	n/a	+	
UK	+	−		Loss

Table 4.8. Summary of indicator trends—non-ferrous metals

Country	Import penetration	Relative production	Export intensity	Overall
DE			+	
DK				
FI		+	−	
NL		+		
SE	+	+	+	Gain
SI		n/a		
UK	+	−	+	Loss

weak decision rule, as it allows the third indicator to be contradictory. A stronger rule would require that the third indicator be consistent or neutral (i.e. not show any trend). The implications of using this stronger decision rule are discussed in the concluding section of the chapter.

The picture for *meat and meat products* (NACE 15.1) is mixed—both across countries and across the three indicators. For Germany, a declining import penetration and a rising export intensity indicate that there was a gain in competitiveness. In contrast, the increasing import penetration for the United Kingdom combined with a falling export intensity, indicate that competitiveness declined. For Denmark, the increase in import penetration and fall in relative production suggest a decline in competitiveness, although the country did experience an increase in export penetration. For the other four countries, no clear picture emerges, with the indicators either exhibiting contradictory trends, or no significant trends.

The picture is also mixed for *paper and paper products* (NACE 21.2). Increasing import penetration and falling relative production suggest that there was a loss of competitiveness for Germany, The Netherlands, and the United Kingdom. However, in all three cases, the rising export intensity gives a contradictory signal. For Sweden, the declining import penetration and rising relative production trends suggest an improvement in competitiveness, although it did suffer a decline in export intensity over the period. It is not possible to draw any conclusions regarding changes in competitiveness for the other three countries, with each having two contradictory indicators.

For *basic chemicals* (NACE 24.1), there is little evidence of any changes in the two trade-related indicators, with only import penetration for Germany and export intensity for Finland showing significant trends. In contrast, trends in relative production are significant for all six of the countries with available data. However, as noted above, caution should be exercised in placing too much emphasis on this indicator, as the rest of the world (i.e. non-EU25) accounts for a relatively large proportion of trade—in both directions—for the ETR countries in this sector. While increasing import penetration and falling relative production indicate that Germany suffered a loss in competitiveness, Finland gained competitiveness, with both relative production and export intensity increasing.

In contrast to the basic chemicals sector, the trends in the two trade-related indicators for *pharmaceuticals* (NACE 24.4) were significant for all of the countries except Sweden. Increasing import penetration and falling relative production suggest that there was a loss of competitiveness for Germany and Finland, although in both cases export intensity rose over the period. Denmark appears to have experienced an increase in competitiveness, with import penetration declining and relative production increasing. While there was also a significant decline in export intensity, the absolute level was very high—in the 90–100 per cent range across the period—and hence it is not clear whether it is above or below the threshold value. Consequently, the indicator has not been used in the overall assessment for Denmark. For the other countries, no clear picture emerges, with Sweden having only one significant indicator and the other three exhibiting contradictory trends.

In the *glass and glass products* sector (NACE 26.1), Germany and Sweden suffered a loss of competitiveness, with relative production falling in both countries. This was accompanied by rising import penetration in Germany and falling export intensity in Sweden. However, in both cases, the remaining indicator shows a contradictory trend. It is not possible to draw any conclusions regarding changes in competitiveness for the other five countries, with the Netherlands, Finland, and the United Kingdom each having only one significant indicator, while Denmark and Slovenia each have two contradictory trends. No import or export data were available for the Netherlands, so it was not possible to construct either of the trade-related indicators.

Two countries experienced an improvement in competitiveness in the *cement, lime, and plaster* sector (NACE 26.5). Germany experienced a reduction in import penetration and a rise in export intensity; Denmark, a fall in import penetration and an increase in relative production. However, again in each case, the remaining indicator shows a contradictory trend. In contrast, rising import penetration and falling relative production indicate a loss of competitiveness for the United Kingdom. Again, no trade data were available for The Netherlands.

In the *ferrous metals* sector (NACE 27.1–3), Germany and Sweden both gained competitiveness, with increasing relative production and export intensity. However, both also experienced increasing import penetration over the period. The United Kingdom was the only country to suffer a loss of competitiveness, with rising import penetration and falling relative production. For Denmark, none of the three indicators exhibits a significant trend; while for the other three countries, there is either only a single significant indicator, or two contradictory indicators.

The picture is not dissimilar for *non-ferrous metals* (NACE 27.4). The United Kingdom suffered a loss of competitiveness, with rising import penetration and falling relative production, although it did experience an increase in export intensity. For Sweden, increasing relative production and export intensity indicate an improvement in competitiveness. However, again the third indicator is contradictory. Relative production increased for the Netherlands, but neither of the other two indicators shows significant trends. Finland has two contradictory indicators; while for Denmark and Slovenia, none of the available indicators is significant.

4.4 Conclusions

In this chapter, the capabilities of four potential 'competitiveness indicators' at the sectoral level have been assessed in the context of a formal theoretical model. As with any exercise of this type, the results and conclusions of the analysis depend on the underlying structure and assumptions of the model. While the model that has been used is relatively general in some respects and captures many of the salient features of the issue, it is clearly a simplification of reality.[16] In particular, it assumes that the firms produce a single homogeneous product with constant unit costs of production; that they all participate in all markets; and that demand for the product is linear in all markets. Consequently, the results of the analysis should be seen as providing guidance regarding the choice and interpretation of indicators to assess changes in a country's sectoral competitiveness, rather than definitive predictions of what would happen to these variables in practice.

Bearing this caveat in mind, there do appear to be differences in the capabilities of the different indicators to represent changes in competitiveness. Provided that the total number of firms in the sector is relatively large (which is likely to be the case in practice), changes in *share of global production* and changes in *import intensity* both provide good indicators of changes in a country's competitiveness. If a country's competitiveness improves, its share of production increases, while the import intensity of its home market declines. If it deteriorates, then the opposite is true and the indicators move in the opposite directions.

Changes in *export intensity* can also provide a good indicator of changes in competitiveness, but only if one can be sure whether export intensity is below or above a critical threshold value. However, for sectors with significant international trade (i.e. in which firms compete in a relatively large number of overseas markets), this threshold value is likely to be high— that is, in excess of 90 per cent. Consequently, for the large majority of cases, export intensity will be below the threshold. In which case, an increase in export intensity indicates an improvement in competitiveness, while a reduction indicates a deterioration.

---

[16] It is consistent with the model underlying the analysis in Chapter 3 regarding the ability of sectors to pass on increases in production costs in price rises. In this model, the proportion of any increase in its unit production cost that a country is able to pass on is (approximately) equal to its share of the total number of firms producing the product in all competing countries.

Finally, changes in *profitability* (as a percentage of sales) do not—in general—provide a good indicator of changes in competitiveness. One cannot even be sure of the relationship between changes in competitiveness and changes in profitability in individual markets. Furthermore, the mix effects and fixed cost effects induced by the change in competitiveness can be in either direction, making it impossible to draw any reliable conclusions about changes in competitiveness from changes in sector profitability.

On the basis of this analysis, the first three indicators were chosen to assess whether there is any evidence that the competitiveness of any of the seven ETR countries deteriorated over the period 1990–2002 in any of the selected industrial sectors. This required that the 'theoretical indicators' from the model be translated into 'practical indicators' for which empirical values could be constructed using the economic and trade data collated in the COMETR data sets for each country. While this was straightforward for the two trade-related indicators, it was more problematic for the share of global production. In practice, it was necessary to compare the countries' changes in real gross output to the average changes across the EU25 as a whole—using an index of production values. As the EU25 accounts for 70–85 per cent of trade for the seven ETR countries in most cases, this is likely to be a reasonable proxy. However, there are a couple of sectors where the rest of the world accounts for a larger proportion of trade and in these cases, the reliability of the indicator may be compromised.

For a variety of possible reasons, the three indicators rarely show a completely consistent picture in practice. Consequently, it was necessary to define a decision rule in order to make any inferences about changes in competitiveness. The rule that has been used for the assessment is that at least two of the indicators should show a consistent picture, and the results of applying this rule are summarized in Table 4.9(a).

Overall, there is little evidence of a systematic deterioration in competitiveness across the seven ETR countries. There is no evidence of any change in competitiveness over the period for 34 of the 56 cases (i.e. 61% of the total). While there appears to have been a loss of competitiveness in 13 cases (i.e. 23%), there was an improvement in nine cases (i.e. 16%). Only in the case of the United Kingdom is there any evidence of a systematic deterioration, with five of the eight sectors suffering a loss of competitiveness. While Germany lost competitiveness in four sectors, it gained competitiveness in three. Both Denmark and Sweden enjoyed more gains in competitiveness than they suffered losses.

97

Table 4.9. Summary of competitiveness changes

Sector	Country						
	DE	DK	FI	NL	SE	SI	UK
(a) Weak indicator consistency requirement							
Meat & meat products	Gain	Loss					Loss
Paper & paper products	Loss			Loss	Gain		Loss
Basic chemicals	Loss		Gain				
Pharmaceuticals	Loss	Gain	Loss				
Glass & glass products	Loss				Loss		
Cement, lime & plaster	Gain	Gain					Loss
Ferrous metals	Gain				Gain		Loss
Non-ferrous metals					Gain		Loss
(b) Strong indicator consistency requirement							
Meat & meat products	Gain						Loss
Paper & paper products							
Basic chemicals	Loss		Gain				
Pharmaceuticals	Loss	Gain					
Glass & glass products							
Cement, lime & plaster							Loss
Ferrous metals							Loss
Non-ferrous metals							

As was noted above, the decision rule used for the assessment is relatively weak, in that it allows the third indicator to be contradictory. Table 4.9(b) shows the impact of using a stronger rule, in which the third indicator is required to be consistent or neutral (i.e. not show any trend). As one would expect, this stronger consistency requirement reduces the number of cases in which competitiveness appears to have changed; with a deterioration of competitiveness now found in only five of the 56 cases (i.e. 9%) and an improvement in three cases (i.e. 5%). The United Kingdom is still the only country to show any evidence of a systematic deterioration, losing competitiveness in three sectors: meat and meat products; cement, lime, and plaster; and ferrous metals. While Germany suffered a loss of competitiveness in the two chemicals sectors, it gained competitiveness in the meat and meat products sector.

The assessment of changes in competitiveness provided in this chapter is by no means perfect. At this level of aggregation (i.e. NACE 3), the data on which it is based show a great deal of volatility from year to year and, in some cases, are of questionable quality. As has been discussed, it has been necessary to use a proxy for one of the indicators and to arbitrarily define a decision rule to combine the three indicators. However, the conclusion from the assessment is unequivocal. With the possible exception of the United Kingdom, there is no evidence of any systematic loss of competitiveness over the period 1990–2002 across the eight sectors considered. For the other six ETR countries, there are only a few cases where competitiveness changed over the period, and in these cases it improved more often than it deteriorated.

While this chapter has assessed whether there is any evidence of changes in competitiveness, it has not considered what the causes of any changes may have been. Even in those cases where there does appear to have been a deterioration in competitiveness, it does not necessarily follow that this was due to the introduction of environmental tax reforms. Indeed, the tax reform in the United Kingdom was not introduced until the very end of the period under consideration and hence is unlikely to have been a significant causal factor in the changes in competitiveness— notwithstanding any possible pre-announcement effects.[17] The causal relationship between the changes in energy taxes arising from the tax reforms and various economic variables is investigated in more detail in the next chapter.

References

Agnolucci, P., Barker, T., and Ekins, P. 2004. 'Hysteresis and energy demand: the announcement effects and the effects of the UK Climate Change Levy'. Working Paper 51. Tyndall Centre for Climate Change Research.

Bernard, J., and Cote, B. 2002. 'The measurement of the energy intensity of manufacturing industries: a principal components analysis'. Discussion Paper 02–31. Resources for the Future, Washington.

Brander, J. A. 1981. 'Intra-industry trade in identical commodities'. *Journal of International Economics*, 11: 1–14.

—— and Krugman, P. 1983. 'A reciprocal dumping model of international trade'. *Journal of International Economics*, 15: 313–23.

Freeman, S. L., Niefer, M., and Roop, J. 1997. 'Measuring industrial energy efficiency: practical issues and problems'. *Energy Policy*, 25/7–9: 703–14.

Worrell, E., Price, L., Martin, N., Farla, J., and Schaeffer, R. 1997. 'Energy intensity in the iron and steel industry: a comparison of physical and economic indicators'. *Energy Policy*, 25/7–9: 727–44.

---

[17] The Climate Change Levy (and associated Climate Change Agreements for energy-intensive sectors) was introduced in April 2001, but was announced in March 1999. Analysis by Agnolucci *et al.* (2004) concludes that there was a 'pre-announcement effect' on firms' behaviour prior to its actual introduction.

5

The Impact of Energy Taxes on Competitiveness: A Panel Regression Study of 56 European Industry Sectors[1]

Martin K. Enevoldsen,[2] *Anders Ryelund,*[3]
and Mikael Skou Andersen[4]

5.1 Introduction

The original Porter hypothesis states that high national environmental standards will encourage domestic industries to innovate and hence improve competitiveness, in particular when the regulatory standards anticipate requirements that will spread internationally (Porter, 1990, 1998). The main reason, according to Porter and van der Linde (1995), is that environmental regulation puts pressure on industry to innovate new and greener products that, in turn, create better demand conditions for the industry. Moreover, environmental standards encourage industries to find less resource-intensive methods of production, thereby counteracting the initial rise in production costs caused by the regulatory demands. The earlier such regulatory pressures are introduced within a given country vis-à-vis other countries, the higher the chance that any innovative

[1] The authors are grateful for comments and suggestions from Paolo Agnolucci, Policy Studies Institute in London, and Natalia Zugravu-Soilita, Center of Economy at the Sorbonne.
[2] Martin Korch Enevoldsen, National Environmental Research Institute, Manager, Deloitte Business Consulting A/S, Denmark.
[3] Anders Ryelund, Administrator, Aarhus University, and Central Region Denmark, Denmark.
[4] Mikael Skou Andersen, Professor, National Environmental Research Institute, Aarhus University, Denmark.

experiments arising from such pressure will lead to a competitive edge.[5]

The critics of the Porter hypothesis reject the argument that environmental regulation should lead firms down more profitable, innovative avenues. If such opportunities existed, they would have been pursued anyway by rational firms, and, in this light, the regulation is just another distortion that may hamper efficient allocation of resources. Hence, the controversy involves intriguing questions on economic rationality and institutional factors which are very difficult to answer a priori. This chapter makes no attempt to resolve the theoretical question. It merely provides some empirical evidence that can be used to indicate if, and to what extent, there is a Porter effect in one special area of environmental regulation. It is recognized that not all environmental regulation will have the desired effect. Porter agrees that traditional environmental regulations have often violated the principles for a positive impact on competitiveness by imposing rigid pollution-abatement technologies, rather than leaving room for adaptation, flexibility, and innovation (Porter, 1991). From this point of view, market-based environmental regulation, including environmental taxes, would be better suited to fulfilling the Porter 'prophecy'. On the other hand, emission taxes (at least those without revenue recycling) introduce an out-of-pocket tax expense to polluting firms on top of the extra abatement costs they experience from their attempts to reduce the tax burden. This brings us to the interesting question whether a Porter effect is in fact associated with environmental taxes and, if so, whether environmental taxes have better or worse effects on competition than environmental standards.

The focus of this chapter is the extent to which energy taxes—via the resulting increase in real energy prices, or in their own right—reduce or enhance industrial competitiveness. From a panel data set covering 56 industry sectors throughout Europe over the period 1990–2003, we estimate how changes in real energy taxes and real energy prices affect, on the one hand, competitiveness measured in terms of unit energy costs and unit wage costs and, on the other hand, economic performance expressed in terms of output (value added). Accordingly, the chapter distinguishes between competitiveness as an economic potential, for example in terms of low unit energy costs, and the effects of that potential, which, for example, could be higher economic output and

---

[5] Although, of course, if regulations are introduced too early, this may cause severe problems for the industry, thus hampering innovative efforts.

exports. If industry experiences significantly lower exports and output as a consequence of a tax-imposed increase in real energy prices, this is a clear indication that the outcome resulted because energy taxes reduced competitiveness. Such findings would give us reason to reject the Porter hypothesis in this specific case.

5.2 Modelling the Porter effects associated with energy taxes

A good theoretical model is required in order to estimate the causal subtleties associated with the possible Porter effect of an environmental tax on energy. According to economic theory, the effect of an energy tax will be exactly the same as the equivalent increase in energy prices. Energy price elasticities with respect to energy consumption and output have been extensively documented using a variety of statistical methods in the energy economics literature (for an overview, see Atkinson and Manning, 1995, plus numerous articles in the *Energy Economics* journal). Panel regression and cointegration analyses have been more successful than older methods at capturing the long-term relationship between energy prices and energy consumption. Typically, the studies report long-term own-price elasticities of industrial energy consumption in the range between -0.3 and -0.6 (Barker *et al.*, 1995). This evidence tells us that energy taxes will have a strong environmental effect in the form of reduced energy consumption and hence less combustion of fossil fuels and lower emissions of air pollution.

But what is the impact of energy taxes on competitiveness and economic performance? Energy is not just some environmental problem, but a major input factor into industrial production. Most evidence indicates that rising energy prices have an adverse impact on economic performance (Longva *et al.*, 1988; Smyth, 1993). The two oil crises during the 1970s speak for themselves (Nasseh and Elyasiani, 1984). Hence, it appears unlikely that energy taxes would be the carrier of a true Porter effect. Indeed, it is hard to believe that energy taxes will make room for so much innovation that it more than offsets the problem of rising input prices. But even if the net effect of an energy tax is a reduction in output, a mitigating Porter effect of substantial size may be involved.

Figure 5.1 shows the basic reasoning. Variables that later appear as dependent variables in the analysis are indicated by rectangular boxes and variables that appear as independent variables, or unobserved intermediary causes, are indicated by oval boxes. A number of relevant

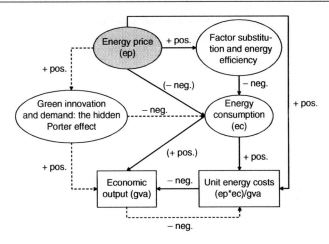

Figure 5.1. A causal model of the Porter effects

independent variables are omitted from the figure, for example government regulation and subsidies to stimulate energy savings, the cost conditions of competitors, etc. The omitted factors are assumed to remain unaltered. There are two separate streams of influences, the first marked by solid lines and the other by dotted.

In the first stream, or chain of effects, the following logic applies. Rising taxes and energy prices will induce firms to substitute towards other input factors (mainly labour and capital), including energy efficiency improvements, which again will lead to lower energy consumption. If the possibilities for innovation and factor substitution are very limited, rising energy prices and taxes may even reduce output since lower energy consumption is not compensated for by other input factors (cf. the relations in parentheses). Factor substitution will, in turn, decrease unit energy costs. On balance, however, unit energy costs are expected to rise because of the higher energy price. The net impact is therefore a reduction in competitiveness on the assumption that competitors (especially foreign competitors) do not experience a similar or higher increase in energy costs. A further implication of increasing unit energy costs (vis-à-vis competitors) is reduced economic output. Overall, the stronger the effect of the mitigating influences in the form of energy savings through factor substitution, the lower the negative impact on economic performance— and the greater the support for the supply-related elements of the Porter idea.

On the other hand, there is a second chain of effects in which rising energy prices and taxes may induce firms to introduce product innovations that minimize the use of resources and other kinds of environmental initiatives that ensure more effective pollution abatement. This may, in turn, stimulate growth either because demand for the specific industrial products increases, or because the initiatives help to create a strong green image, which improves general economic conditions for the firm. This broader green innovation effect is the core of the Porter hypothesis, but it is much more difficult to observe and measure than the first chain of effects.

In the first chain of effects, the Porter element reduces to the mitigating influence that factor substitution has on the original negative economic impact of higher energy prices. One would never expect factor substitution to be so high that unit energy costs actually decline and output grows as a result of higher energy prices and taxes. However, the second chain of effects—the demand-related green innovation effect—introduces the possibility that, on balance, green energy taxes reduce competitiveness and output only slightly or perhaps even lead to improvements.

In the subsequent statistical analyses, we will test whether Porter hypotheses of various degrees are supported by the evidence relating to energy taxes. One of the most interesting questions is, of course, whether the hidden Porter effect is strong enough to offset the expected adverse impact of energy prices on economic performance. Hence, if we find a positive relation between energy taxes, competitiveness, and output, this would indicate the existence of a *radical Porter effect*. This would indeed be contrary to ordinary economic reasoning and move the scope of the Porter hypothesis beyond its usual application to non-fiscal instruments of environmental regulation.

More likely, there is a chance that Porter effects working through the factor substitution channel and the demand-related innovation channel strongly reduce the original negative effects of energy taxes on unit energy costs and output. If that turns out to be the case, it will indicate the existence of a *mitigating Porter effect* even with respect to tax instruments of environmental regulation. Finally, if economic performance is severely harmed by rising energy prices and taxes as assumed by mainstream theory, it indicates the *absence of Porter effects* in this area.

5.3 Data and method

The analysis is based on the COMETR data set covering eight industrial sectors in seven different European countries for the period 1990–2003.

Table 5.1. The industry sectors
in the data set

Sector (NACE 3-digit)

15.1 Meat industry
21.2 Paper and cardboard articles
24.1 Basic chemicals industry
24.4 Pharmaceuticals industry
26.1 Glass industry
26.5 Cement, lime and plaster
27.1–27.3 Basic ferrous metals
27.4 Basic non-ferrous metals

This amounts to a maximum number of 784 observations on each variable. The countries included are Denmark, Finland, Germany, Netherlands, Slovenia, Sweden, and the UK. The sectors included are as shown in Table 5.1.

Accordingly, the data set contains a mixture of energy-intensive (24.1, 26.1, 26.5, 27.1–3, and 27.4) and medium energy–intensive sectors (15.1, 21.2, and 24.4). Data on a large number of energy-related and economic variables has been collected for each of the industry sectors by teams in the respective countries. In Table 5.1, we provide a list of the subset of variables that we included in the panel regressions. All economic variables are in fixed 2000 prices.

From the causal model in Figure 5.1, we note that endogeneity problems apply to the set of variables that, as a minimum, would be required to estimate both unit energy costs and economic output. Unit energy costs are influenced by economic output as economies of scale give rise to less energy use per unit when production increases and, in reverse, economic output is influenced by unit energy costs as competitiveness influences the output level.

In deciding about the appropriate statistical methods to use in estimating the causal relations, we were limited by data availability. It was not feasible to extend the time series beyond the 14-year period 1990–2003. This relatively short time series ruled out the application of VAR and cointegration techniques, which would have been preferred (in combination with panel data techniques), given the challenge of endogenous variables, the supposed dynamic character of the interrelations, and the often reported cointegrating nature of the central variables (energy prices, energy consumption, and output).[6]

---

[6] See e.g Hunt and Manning (1989); Hunt and Lynk (1992); Bentzen and Engsted (1993); Barker *et al.* (1995); Asafu-Adjaye (2000); Stern (2000); Enevoldsen (2005: 187–220).

Table 5.2. List of variables applied in the panel regression

Variable	Description
gva	Gross value added (€ in fixed 2000 prices). Deflated by the producer price index (PPI),[1] GVA measures real economic output and is also used as a proxy measure of industrial production volume in economic terms.
yvol	The value of total industrial output (€ in fixed 2000 prices). It is used as a proxy measure of industrial production volume in physical terms.
encon	Total energy consumption (GJ).
uec	Unit energy costs. Total energy costs (€) per value added (€). Total energy costs are divided by GVA.
ulc	Unit labour costs. Labour costs (€) per value added (€). Total compensation of employees is divided by GVA.
urc	Unit raw materials costs. Total intermediate consumption (€) exclusive energy costs per value added (€).
uic	Unit input costs. Total factor input costs (€) per value added (€).
ep	Real energy price (€ in fixed 2000 prices). Total energy costs are divided by total energy consumption and thereafter deflated by PPI.
epex	Real energy price exclusive taxes (€ in fixed 2000 prices). Total energy costs exclusive taxes are divided by total energy consumption and thereafter deflated by PPI.
etax	Real energy taxes (€ in fixed 2000 prices). Total energy taxes are divided by total energy consumption and thereafter deflated by PPI.
wage	Real wage (€ in fixed 2000 prices). Total compensation of employees is divided by the total number of employees and thereafter deflated by PPI.

Note: [1] The producer price index (PPI) for each sector is used as a substitute for the sector-specific GDP deflator (the price level of all input factors), for which data were not available.

Yet, with the available data, panel regression techniques were indeed feasible. When the data set contains not only cross-sections[7] but also repeated observations over time for each cross-section, panel regression techniques may provide better estimates compared to disjointed ordinary least square (OLS) regressions of each individual cross-section. This is because panel regression takes into account the variance across sections (and time) in making the estimates. However, panel regression is appropriate only if it makes sense to pool observations to search for some joint coefficient estimates, while still allowing for certain differences between individual sectors and/or time periods. In our case, we have a panel data structure, where the cross-sections (i.e. groups) are the respective industry sectors for which data were collected, and the time series are the annual observations between 1990 and 2003 for each industry sector.

Since all the chosen industry sectors are characterized as energy intensive or medium energy intensive, and all of them reside in countries that

[7] Cross-sections refer to observations across different individuals, sectors, or countries at some point in time.

Table 5.3. Unit energy costs by industrial (NACE) sector and country—average energy costs (€) per 100€ value added

	Denmark	Finland	Germany	NL	Slovenia	Sweden	UK
15.1	5.0	4.9	6.9	4.9	8.6	3.9	5.8
21.2	3.2	4.2	5.8	4.5	15.0	4.7	6.8
24.1	10.6	37.3	25.3	20.7	24.0	17.5	28.6
24.4	2.3	3.5	2.7	4.2	3.5	1.6	2.5
26.1	7.0	14.4	15.7	13.6	23.4	13.0	8.2
26.5	30.0	37.0	42.0	9.5	64.5	38.6	25.0
27.1–3	11.4	47.1	32.5	24.0	72.0	28.7	47.7
27.4	4.5	28.6	26.0	33.6	188.6	27.5	19.1

count as advanced North European economies (with the exception of Slovenia), it is assumed that the data set is sufficiently homogeneous to pool the observations.[8] In Table 5.3, average unit energy costs are shown for each cross-section to provide an idea about the homogeneity across sections with respect to one of the most central variables in the analysis.

Tests were carried out to determine the appropriate extent of pooling and on the basis of these tests it was decided to use panel regression methods that allow the individual effects to differ across sectors, but not over time.[9]

The panel regression analyses centre around two basic models:

$$y_{it} = a'_i + \beta' x'_{it} + u_{it} \tag{5.1a}$$
$$y_{it} = \beta' x'_{it} + (a + u_i + \epsilon_{it}) \tag{5.1b}$$

The first model is the fixed effects panel regression model. In this model, the omitted sector specific structural variables are treated as fixed constants over time (a'_i). The second is the random effects model in which the individual effect is considered as a time-invariant component in the error term, that is, a random disturbance (u_i) of the mean unobserved heterogeneity (a). Although both models incorporate individual

[8] Excluding Slovenia from the data set does not change the findings presented here.
[9] We tested for the existence of individual group effects and time effects by analysing the variance using the *pstats* procedure in RATS. The method works by decomposing the variance into three different alternatives, one with a random component plus individual effects only, a second with a random component plus time effects only, and a third with random plus joint individual and time effects. F-tests from one-factor and two-factor analyses of the variance are calculated for the three alternatives. The test results showed that individual effects alone were highly significant, that time effects alone were not significant, and that joint effects were also significant. But the test results also showed that joining the effects adds very little to model perfection as compared with the individual effects model (which has the advantage of leaving many more degrees of freedom). Therefore, the individual effects model was selected as our general approach to pooling the data.

effects stemming from omitted variables, the central difference is that the random effects model represents the individual effects by a random component in the error term and thus prohibits correlation between these individual effects and the regressor variables x'.

Because of the endogeneity problems that apply to the models under investigation (see above), there is most likely correlation between the residuals and regressors and hence it is not very likely that the individual effects stemming from omitted variables are uncorrelated with the independent variables. This suggests that we use the fixed effects model.[10] Hausman specification tests were carried out to verify that the fixed effects model is superior to the random effects model for the relations we want to estimate.

5.4 The relation between energy taxes, competitiveness, and output

There are a variety of indicators for industrial competitiveness. Focusing on the impact of energy taxes, the most relevant measure of price competitiveness is *unit energy costs*, which is defined as total energy expenditure (including taxes) per unit of gross value added in market prices. While unit energy costs are a partial measure of the price competitiveness of an industry, it is also a measure of energy intensity. Hence, if unit energy costs decrease as a consequence of substitution of labour for energy, which then turns out to increase unit labour costs, the firm will, on balance, not necessarily become more price competitive, but it will surely be less energy intensive. However, if real unit energy costs decrease and other unit input costs remain stable, it is indeed an indication that price competitiveness improved. We therefore investigated the impact on two partial measures of price competitiveness: unit energy costs and labour unit costs (defined as total wages and compensation per unit of gross value added in market prices).[11]

The original single equations used for estimating unit energy costs and unit labour costs are listed as equations (5.2) and (5.3) below: all variables in these and the coming equations refer to their logarithmic

---

[10] See Hsiao (2003: 41 ff.) for further discussion of the theoretical and methodological considerations in choosing between fixed and random effects models.

[11] It would be relevant to investigate the impact on total unit input costs (including costs of capital and raw materials) also, but since the data set does not contain sufficient information on these costs, it was not feasible to use it as an independent variable in a separate estimation.

(ln) values to make the results interpretable in percentage elasticities. Equations 5.2 and 5.3 appear as fixed effect models, where the a_i and D_i are the fixed effect constants for each individual sector, $\tau(t)$ is a general linear trend, and $\mu_{it}(u_{it})$ are the residuals. The right-hand side includes a lag of the dependent variables. The remaining symbols represent estimates of the regressor coefficients that are assumed to be joint for all sectors. In the underlying work, this assumption was relaxed by carrying out individual tests at sector level using the same basic model, which in this more disaggregated setting allows coefficients to vary across industry sectors, or across countries.

Using Hausman specification tests,[12] it was investigated whether random effects models were more appropriate for estimating the unit cost equations and in both cases the answer was negative, as we already suspected given the endogeneity problem.

$$
\begin{aligned}
\text{uec}_{it} &= a_i + \beta^* \text{epex}_{it} + \chi^* \text{etax}_{it} + v^* \text{ulc}_{it} + o^* \text{urc}_{it} + \delta^* \text{gva}_{it} \\
&\quad + \tau^* \text{trend} + \phi^* \text{uec}_{i,t-1} + \mu_{it} \quad\quad (5.2)
\end{aligned}
$$

$$
\begin{aligned}
\text{ulc}_{it} &= D_i + w^* \text{wage}_{it} + e^* \text{uec}_{it} + r^* \text{urc}_{it} + y^* \text{gva}_{it} + t^* \text{trend} \\
&\quad + f^* \text{ulc}_{i,t-1} + u_{it} \quad\quad (5.3)
\end{aligned}
$$

The assumption behind the basic models is that unit energy and unit labour costs are, of course, determined first and foremost by the real price of energy and labour respectively. Moreover, they are determined by the unit costs of other input factors. For example, increasing unit labour costs will probably encourage industrial firms to use more energy as a substitute and thus raise unit energy costs. Unit costs are also influenced by the output quantity (gva) as economies of scale reduce average production costs and since growth tends to reduce problems of overcapacity.

Our proxy measure for unit raw material costs (cf. Table 5.2) is subject to more uncertainties than our similar measure for unit energy and labour costs. Moreover, it is not entirely clear that increasing raw material costs would lead to factor substitution towards energy, as the consumption of the two often go together. We therefore tested the possibility of excluding urc as a regressor from both equations (5.2) and (5.3) and found that it

---

[12] In order to harmonize the number of coefficients and covariance matrix from the two competing models, and thus simplify the calculations involved in the Hausman test, a general constant was added to the fixed effects model. The constant creates no disturbance, as it washes out in the performed regression.

Table 5.4. Unit energy costs—equation (5.2) estimated with fixed effects and robust errors

Variable	Parameter estimate	Standard error	t-stat	p-value
epex	$\beta = 0.527$	0.0678	7.78	0.000
etax	$\chi = 0.030$	0.0071	4.27	0.000
uwc	$\nu = 0.293$	0.0581	5.05	0.000
urc	excluded			
gva	$\delta = -0.511$	0.0483	-10.57	0.000
trend	$\tau = 0.005$	0.0021	2.62	0.008
uec(t−1)	$f = 0.241$	0.0457	5.29	0.000

Table 5.5. Unit labour costs—equation (5.3) estimated with fixed effects and robust errors

Variable	Coefficient estimate	Standard error	t-stat	p-value
wage	$w = 0.343$	0.0360	9.55	0.000
uec	$e = 0.123$	0.0259	4.73	0.000
urc	$r = 0.145$	0.0277	5.25	0.000
gva	$y = -0.325$	0.0385	-8.45	0.000
trend	$t = -0.004$	0.0019	-2.41	0.016
ulc(t−1)	$\Phi = 0.330$	0.0393	8.41	0.000

could be excluded from the former, but not the latter.[13] Subsequently, we estimated the uec equation without the urc variable.

Tables 5.4 and 5.5 show the single equation estimation of equations (5.2) and (5.3) without the urc regressor in equation (5.2). The 56 dummy coefficients accounting for fixed effects are not reported in the tables. The model statistics show a very good fit for both the uec and the ulc equations ($R^2 = 0.989$ and 0.968, respectively). The equations were estimated with *robust errors* option in the RATS software package in order to correct the covariance matrix to allow for complex residual behaviour, including heteroscedasticity and serial correlation. The estimated models were also tested for heteroscedasticity by means of the White test (1980) and for serial correlation by the Breusch-Godfrey test and the tests could not confirm the null hypothesis of respectively homoscedasticity and no autocorrelation among the lagged residuals. When estimating the models

[13] The test was carried out with the *exclude* command in RATS which provide F-statistics, or in this case Chi-square (because robust errors were used), for the restriction that the listed coefficients are zero.

without the lagged dependent variables, White and Durbin–Watson tests indicated similar problems.

The problems may relate to the many dummy variables included, but it could also be due to the endogeneity of the gva, uec, and ulc variables which, in any case, suggests that it is preferable to estimate the uec and ulc equations simultaneously, along with an output equation, that is, as a three-equation system. Before we move on to this next step, and before we start to interpret the results, we will briefly discuss and provide a first single-equation estimate of output (gva).

The central measure of economic performance is growth in terms of output. Gross value added is the normal indicator of economic growth, and we therefore investigated the impact of energy prices, energy taxes, labour costs, and raw materials costs on value added. According to economic theory, industrial supply is influenced by input factor prices. If the cost of production factors go up, the cost of supplying the same quantity will increase and hence supply will be reduced, causing a decline in output *ceteris paribus*. It is the total marginal costs of input factors that determine supply and hence it should not matter whether higher costs are caused by higher energy costs, labour costs, or raw materials costs. Furthermore, if an increase in one of these costs is fully offset by decline in one or more of the other cost factors, supply should not be affected.

Output is also influenced by demand, that is, the consumer's willingness to pay for the products. Ideally, output should therefore be estimated by the means of a simultaneous supply and demand equation. However, in our case, we do not have sufficient information to estimate demand. Yet, the output measure is, to a certain extent, corrected for the demand factor, as it is deflated by the producer price index (PPI). For our purposes, it should therefore be sufficient to estimate a supply-focused output equation:

$$\text{gva}_{it} = \kappa_i + \gamma^*(\text{uec}_{it} + \text{ulc}_{it} + \text{urc}_{it}) + s^*\text{trend} + \psi^*\text{gva}_{i,t-1} + \epsilon_{it} \qquad (5.4)$$

In the output equation, unit input costs are represented by uec + ulc + urc, which should cover the full input costs since unit raw materials costs (urc) are measured here as all intermediary costs of production, excluding energy costs and compensation of employees.[14]

Model 5.4 has a very high $R^2(= 0.996)$, and the coefficients all have the expected sign, just like the coefficients in models (5.2) and (5.3),

---

[14] It therefore includes intermediary costs related to administration also.

Table 5.6. Gross value added—equation (5.4) estimated with fixed effects and robust errors

Variable	Coefficient	Standard error	t-stat	p-value
unit input costs	$\gamma = -0.490$	0.0380	−12.89	0.000
trend	$\zeta = 0.013$	0.0017	7.98	0.000
gva(t−1)	$\psi = 0.544$	0.0438	12.42	0.000

but again there are problems with endogeneity, heteroscedasticity, and autocorrelation. In the next step, we therefore specify a full system of simultaneous equations—especially with a view to getting a clearer picture of the reciprocal influence between output, unit energy costs, and unit labour costs.

From the observation that the trend variable is more important in the output equation than in the equations (5.2) and (5.3) (cf. the higher t-statistic for the trend in Table 5.6 versus Tables 5.4 and 5.5), we made further investigations and came to the conclusion that the output model in the full equation system could be improved by working with sector-specific trends instead of a common trend. We therefore added a fixed effects dummy trend variable, but only to the output equation within the system.[15]

$$gva_{it} = \kappa_i + \gamma^*(uec_{it} + ulc_{it} + urc_{it}) + s_i^*trend + \psi^*gva_{i,t-1} + \epsilon_{it} \quad (5.5a)$$

$$uec_{it} = a_i + \beta^*epex_{it} + \chi^*etax_{it} + \nu^*ulc_{it} + \delta^*gva_{it} + \tau^*trend$$
$$+ \phi^*uec_{i,t-1} + \mu_{it} \quad (5.5b)$$

$$ulc_{it} = D_i + w^*wage_{it} + e^*uec_{it} + r^*urc_{it} + y^*gva_{it} + t^*trend$$
$$+ f^*ulc_{i,t-1} + u_{it} \quad (5.5c)$$

Equations (5.5a) to (5.5b) were estimated with the *nlsystem* procedure in RATS, which allows us to work with complex simultaneous equations, and use a generalized method of moments (GMM) estimator. GMM estimators apply an optimal weighting matrix to the orthogonality conditions that are used for correcting the covariance matrix (Hansen, 1982). The applied GMM estimator corrects, as much as possible without changing the model, for problems with heteroscedasticity and serial correlation. Moreover, simultaneous estimation allow us to work with endogenous

---

[15] It might have been relevant to work with sector-specific trends also for the uec and ulc equation, but that would require the estimation of another 112 parameters and thus deplete our degrees of freedom to an unacceptable extent. The dummy trend vector was therefore used where it mattered most—in the output equation.

Table 5.7. Simultaneous estimation of gva, uec, and ulc—equations (5.5a)–(5.5c) subject to non-linear GMM estimation

Equation	Variable	Coefficient	Std. error	t-stat	p-value
GVA	unit input costs	$\gamma = -0.241$	0.0123	19.66	0.000
GVA	gva(t−1)	$\psi = 0.200$	0.0283	7.08	0.000
UEC	epex	$\beta = 0.546$	0.0494	11.04	0.000
UEC	etax	$\chi = 0.021$	0.0079	2.66	0.008
UEC	uwc	$\nu = 0.066$	0.0699	0.95	0.344
UEC	gva	$\delta = -0.534$	0.0585	−9.11	0.000
UEC	trend	$\tau = 0.009$	0.0023	3.74	0.008
UEC	uec(t−1)	$\Phi = 0.289$	0.0317	9.12	0.000
ULC	wage	$w = 0.372$	0.0365	9.55	0.000
ULC	uec	$e = 0.050$	0.0313	1.60	0.109
ULC	urc	$r = 0.164$	0.0246	6.67	0.000
ULC	gva	$y = -0.265$	0.0401	−6.61	0.000
ULC	trend	$t = -0.006$	0.0017	−3.54	0.016
ULC	ulc(t−1)	$f = 0.362$	0.0314	11.51	0.000

variables (in this case, gva, uec, and ulc) vis-à-vis instrumental variables (the regressors that appear only on the right-hand side) and thus with a theoretically more adequate model. In such a model, the problems with residual variance and residual correlation are expected to be smaller.

The cost of simultaneous equations is the loss in degrees of freedom when so many parameters have to be estimated at once. Out of 783 observations, 435 were usable (the rest were skipped due to missing data in some variables). In total, 238 parameters had to be estimated, including 224 dummy variables! That still leaves enough degrees of freedom to be confident about the estimates. The R^2 for the respective equations within the system are, as expected, very similar to those of the single equations. Yet, most of the coefficient estimates are quite different, as we see from Table 5.7.

5.5 Interpretation of results

With the final estimation of the major dependent variables from equations (5.5a)–(5.5c), we can go on to interpret the results.

5.5.1 Unit energy costs

The results show, as expected, that rising energy prices over time lead to increasing unit energy costs, although the impact is not a one-to-one relation. From the estimation of the simultaneous equation (5.5b), it appears that the long-term impact—after factor substitution, output adjustment, etc.—of a 1 per cent increase in the real energy price is a 0.77 per cent increase in unit energy costs.[16] This is very close to the estimate in the single uec equation (cf. Table 5.4).

More interestingly, the effect on unit energy costs of a 1 per cent energy tax increase is 26 times as little (0.546/0.021) compared to a 1 per cent increase in the market energy price. Since the level of market energy prices is, on average, 17 times higher than energy taxes for the observations in this data set, the result indicates that a change in the energy tax has a relatively lower effect on unit energy costs than the same absolute change in the market energy price. Hence, there is some indication that energy taxes do not harm competitiveness as much as ordinary price increases.[17] The total long-term effect of a 1 per cent energy tax increase is that unit energy costs go up by some 0.03 per cent.

Unit labour costs (ulc) tend to have a weak positive impact on unit energy costs, which is what we would expect from factor substitution. Yet, the estimate is only significant in the single equation. The results, moreover, show that higher output reduces unit energy costs, which is also expected, although it is a bit unexpected that the relation is almost as strong as the energy price effect. This might indicate that the real recursive relation between gva and uec is not fully captured, even in the simultaneous equation system.

5.5.2 Unit labour costs

The results show the same basic pattern as above. Unit labour costs are first of all determined by the price of labour, that is, real wages. But the wage–ulc relation is more inelastic than the epex–uec relation. Unit energy costs and unit raw materials costs both have a positive influence on unit labour costs, as firms substitute towards labour—especially when the price of raw materials go up. And again, output works through economies of scale to reduce unit labour costs.

---

[16] This is because $0.546/(1 - 0.289) = 0.77$ after taking into account the correction for lagged dependent variables (cf. Greene, 2003: 727).

[17] In the single equation, the result is different. Here energy taxes tend to have the same effect as market energy prices when the same absolute size is compared. Yet the simultaneous estimation is probably more credible, as it takes into account the recursive impact from energy taxes via output.

5.5.3 Economic output

The output equation clearly illustrates the need for simultaneous estimation. The single equation estimate indicates an extremely steep supply
curve since a 1 per cent increase in unit input costs leads to a 1.07 per cent
decline in output (after correcting for the lagged endogeneous variable).
The estimate from the simultaneous equation (5.5a) is theoretically more
justifiable and also much more realistic. According to this estimate, a 1
per cent increase in unit input costs leads to a 0.3 per cent decline in
output.

5.5.4 The effects of energy taxes on economic performance

On that basis, we conclude as follows with respect to the average impact
of energy taxes on competitiveness. Competitiveness is reduced as a consequence of higher energy prices, since it leads to both higher unit energy
costs and unit labour costs. However, unit energy costs only go up by 0.3
per cent and unit labour costs by 0.023 per cent if energy taxes increase
by as much as 10 per cent. If, for example, energy costs amount to 10 per
cent and labour costs amount to 50 per cent of all input costs, the final
effect of a 10 per cent energy tax increase will be a small 0.04 per cent
decline in output. Hence competitiveness and economic output is not
affected very much by changes in energy taxes. This conclusion applies to
changes within the scope of fluctuations experienced in the period under
investigation. Moreover, it does not distinguish between the tax level at
which the tax increase occurs. Higher energy tax increases may thus have
more drastic (exponential) effects, especially if introduced from an already
high tax level.

5.5.5 Searching for the Porter effects

In the theoretical section, we identified two possible Porter effects, a
supply-oriented effect that mainly operates via factor substitution and
energy efficiency improvements and a demand-oriented effect mainly
operating via green innovation that raises demand for industry products.
In other words, the first Porter effect mainly works by reducing energy
consumption and the second mainly works by increasing the consumer's
willingness to pay. The influence of market energy prices and taxes on
energy consumption can be roughly approximated by the following single

equation:

$$\text{encon}_{it} = D_i + a^*\text{epex}_{it} + b^*\text{etax}_{it} + c^*\text{wage}_{it} + d^*\text{gva}_{it} + g^*\text{trend} + h^*\text{encon}_{i,t-1} + u_{it} \tag{5.6}$$

A more correct estimate of energy consumption would be expected from simultaneous factor input equations, but since we have no reliable data on the price of raw materials and capital, we settle for the proxy type in equation (5.6), which normally works reasonably well in estimating energy consumption.

The results show that the long-term elasticity of energy consumption with respect to market energy prices is –0.435, which is well in accordance with recent findings in the area of industrial energy price elasticities.[18] Industrial output quantity has the expected positive impact on energy consumption, although it is far from constant returns to scale.

It could be the case that energy taxes mainly work through the demand-related Porter effect on output, which therefore implies a positive recursive influence on energy consumption via output. To test the idea that energy taxes have a positive direct impact on demand, we re-estimated the simultaneous equation system (5.5a–c) by adding the etax variable to the right-hand side of equation (5.5a). Although this implies some multicollinearity, the problem should be very small, as the energy tax is only a tiny part of total input costs. The results are shown in Table 5.9.

We find that energy taxes have a very significant direct impact on output, in that a 10 per cent increase in energy taxes leads, on average, to an increase in gva by some 0.23 per cent. The other two coefficients and their statistics remain relatively stable after the inclusion of etax in equation (5.5a). Although the additional results in Table 5.8 and 5.9 are very far from answering all open questions related to the hidden Porter effect,[19] we have at least provided an indication that there is indeed a Porter effect that mitigates the immediate negative impact of green energy taxes on economic performance. We also reach the tentative conclusion that the Porter effect works through demand-related green innovation rather than supply-related factor substitution.

---

[18] We choose to exclude the lagged dependent variable this time as it tends to overdetermine the regression. The main conclusions are not affected by whether it is included or not, although the long-term coefficients tend to moderate. Yet, heteroscedasticity and autocorrelation problems apply (the Durbin-Watson statistic is only 1.09). A truly dynamic cointegration model would probably be required to do away with this and would perhaps be able to give a better account of the tax effect.

[19] A direct regression of willingness-to-pay (demand) against energy taxes and other demand-related variables would have been preferable, but is not feasible with the available data set.

Table 5.8. Energy consumption—equation (5.6) estimated with fixed effects and robust errors

Variable	Coefficient estimate	Standard error	t-stat	p-value
epex	a = −0.435	0.0641	−6.78	0.000
etax	b = 0.011	0.0081	1.35	0.178
wage	c = −0.093	0.0723	−1.29	0.198
yvol	d = 0.335	0.0443	7.56	0.000
trend	h = 0.004	0.0029	1.62	0.105

Table 5.9. Simultaneous estimation of gva, uec and ulc—re-estimation of (5.5a)–(5.5c) by adding etax with coefficient named π to (5.5a)

Equation	Variable	Coefficient	Std. error	t-stat	p-value
GVA	etax	$\pi = 0.023$	0.0055	4.24	0.000
GVA	unit input costs	$\gamma = -0.241$	0.0120	−20.05	0.000
GVA	gva(t − 1)	$\psi = 0.206$	0.0277	7.44	0.000

5.6 Conclusions

At the beginning of the chapter, we posed the question whether Porter effects, which are normally associated with environmental regulation of a more traditional kind, also play a role with respect to economic instruments of environmental regulation, such as (green) energy taxes. In general, the literature has experienced difficulties in providing clear-cut evidence in favour of the Porter hypothesis. Yet economic instruments of environmental regulation have quantitative properties that make it easier to test for effects on competitiveness and economic performance. In this chapter such an attempt was made with respect to energy taxes. Energy taxes were described and carefully measured, along with a number of other central economic variables in the data set containing time series of eight relatively energy-intensive industry sectors in seven different countries.

By means of econometric panel regression techniques, we have demonstrated the impact of market energy prices, energy taxes, labour and raw materials costs on price competitiveness and economic output. We have quantified the economic impact of energy taxes and have shown that higher energy taxes lead to a moderate increase in unit energy costs and a small increase in total unit input costs, which again lead to an even smaller reduction in economic output according to our simultaneous equation model. We have also demonstrated that, with a high probability, the very moderate negative economic impact is the result of Porter effects—in particular because the application of (mainly green) energy taxes stimulates efforts within the industries that in turn raises demand for their products and thus has a direct positive impact on output that counteracts the negative supply effects of the tax increase. We also provided strong indications that energy taxes have different effects on competitiveness and output than market energy prices of a similar size.

With the available data, it is, however, difficult to say whether the interesting effects can be ascribed solely to energy taxes, or if energy taxes go hand in hand with various kinds of government support in a systematic way (e.g. earmarked subsidies for energy-savings, public information and marketing campaigns, and compensation of industries with respect to other taxes or social contributions). More rigorous testing would require some measure of government support to be included in the models. It would also require a better demand model than the proxy we have devised under the present conditions, along with more reliable data on capital and the price of raw materials. Moreover, it would require much longer time series that allow for dynamic VAR estimation methods and hence a more reliable account of the complex endogeneity among the central variables.

References

Asafu-Adjaye, J. 2000. 'The relationship between energy consumption, energy prices and energy growth: time series evidence from Asian developing countries'. *Energy Economics*, 22/6, Dec.: 615–25.

Atkinson, J., and Manning, N. 1995. 'A survey of international energy elasticities', in Barker *et al.* (1995), 47–105.

Barker, T., Ekins, P., and Johnstone, N. (eds.) 1995. *Global Warming and Energy Demand*. London and New York: Routledge.

Bentzen, J., and Engsted, T. 1993. 'Short- and long-run elasticities in energy demand'. *Energy Economics*, 15/1: 9–16.

Enevoldsen, M. 2005. *The Theory of Environmental Agreements and Taxes: CO_2 Policy Performance in Comparative Perspective*. Cheltenham: Edward Elgar.

Greene, W. H. 2003, *Econometric Analysis*, 5th edn. Upper Saddle River, NJ: Pearson Education, Prentice Hall.

Hansen, J. A. 1982. 'Large sample properties of generalized method of moments estimators'. *Econometrica*, 50: 1029–54.

Hsiao, C. 2003. *Analysis of Panel Data*, 2nd edn. Cambridge: Cambridge University Press.

Hunt, L. C., and Lynk, E. L. 1992. 'Industrial energy demand in the UK: a co-integration approach', in D. Hawdon (ed.), *Energy Demand: Evidence and Expectations*. Guildford: Surrey University Press.

—— and Manning, N. 1989. 'Energy price- and income-elasticities of demand: some estimates for the UK using the co-integration procedure'. *Scottish Journal of Political Economy*, 36/ 2: 183–93.

Longva, S., Olsen, Ø., and Strøm, S. 1988. 'Total elasticities of energy demand analysed within a general equilibrium model'. *Energy Economics*, 10/4, Oct.: 298–308.

Nasseh, A. R., and Elyasiani, E. 1984. 'Energy price shocks in the 1970s: impact on industrialized economies'. *Energy Economics*, 6/4, Oct.: 231–44.

Porter, M.E. 1990: *The Competitive Advantage of Nations*, London: MacMillan.

—— 1991. 'America's green strategy'. *Scientific American, 264*: 168.

—— 1998. *On Competition*. Boston: Harvard Business School Press.

—— and van der Linde, C. 1995. 'Toward a new conception of the environment-competitiveness relationship'. *Journal of Economic Perspectives*, 9/4: 97–118.

Smyth, D. J. 1993. 'Energy prices and the aggregate production function'. *Energy Economics*, 15/2, Apr.: 105–10.

Stern, D. I. 2000. 'A multivariate cointegration analysis of the role of energy in the US macroeconomy'. *Energy Economics*, 22: 267–83.

White, H. 1980. 'A heteroscedasticity-consistent covariance matrix estimator and direct test for heteroscedasticity'. *Econometrica*, 48: 817–38.

6

Energy-Intensive Industries: Approaches to Mitigation and Compensation

Mikael Skou Andersen[1] and Stefan Speck[2]

6.1 Introduction

There are quite different arrangements for mitigation and compensation in place with respect to energy-intensive industries in Europe's ETR countries. The differences in scope are interesting to study and contrast, as they reflect somewhat different strategies for safeguarding competitiveness. Under EU state aid rules, a common legal framework has gradually emerged which constrains the options of member states, but by no means ensures full harmonization. The aim of this chapter is to provide an overview and analysis of the approaches developed and in place.

Conceptually, the OECD (2001) distinguishes between *ex-ante mitigation* and *ex-post compensation*. Ex-ante mitigation includes purposeful modifications of the *tax base*, omitting certain fuels for selected emitters, as well as selective reductions in *tax rates* for certain emitters, possibly in a phased way, with thresholds above which tax rates are reduced or capped. Such reductions can be contingent in part on *agreements* to undertake energy efficiency measures. Ex-post compensation includes *subsidies* offered to support specific industries. More important ex-post is *revenue recycling,*

[1] Mikael Skou Andersen, Professor, Department of Policy Analysis, National Environmental Research Institute, Aarhus University, Denmark.
[2] Stefan Speck, Senior Economist, Department of Policy Analysis, National Environmental Research Institute, Aarhus University, Denmark and Kommunalkredit Public Consulting, Austria.

which can take place either via a lowering of social security contributions or via a lowering of income taxes.[3]

The OECD notes that '[m]itigation measures reduce the environmental effectiveness of the tax by cancelling out some of the incentives to change consumption and investment behaviour' (OECD, 2001: 29). However, where ex-post compensation provides for complementary incentives, environmental results may still be attained. Voluntary sectoral agreements, for instance, commit companies to deliver a similar amount of CO_2 reductions as would follow from purely economic incentives. The recycling of tax revenue, for example for targeted subsidies to energy efficiency measures, may help lower the marginal cost curve and hence mitigate the need for more substantive tax rates.

6.2 Ex-ante mitigation: tax-base modifications and reductions in tax rates

6.2.1 Sweden

Sweden's 1989 environmental tax reform transformed the existing practice of energy taxation on industry to a combined energy and CO_2 tax base, which resulted in a level of carbon-energy taxation that by international standards was exceptionally high. In the years 1991 and 1992, energy-intensive industries were charged effective carbon-energy tax rates of EUR 40–50 per tonne of CO_2 or EUR 7–12 per GJ. The tax rates for industry were set in line with those for households. Overall, the initial tax rates corresponded to the European Commission's 1990 proposal for a carbon-energy tax of US$ 10 per barrel of oil.

The initial scheme was soon overturned by competitiveness concerns, and the CO_2 tax rate for all industrial sectors was reduced to 25 per cent of its initial level, while the traditional energy taxation on industry that had been in place since 1974 was abolished. In 2006, industry paid 21 per cent of the CO_2 tax rate levied on households. From 1993 and until the implementation of the Energy Taxation Directive (European Commission, 2003), Swedish industry was also exempt from electricity taxation.

---

[3] Social security contributions are in some countries regarded as charges rather than as taxes; however, for the sake of simplicity, we will discuss all compulsory payments as part of the overall tax burden on labour.

The initial scheme as well as its successor allowed a cap on carbon-energy taxation to sectors where energy taxation exceeded 1.7 per cent of the value of sales.[4] The mitigation approach was to some extent discretionary in that companies were required to apply individually in order to obtain the cap. Among the concerns already highlighted by a government investigator in 1991 (SOU, 1991: 90), as Sweden was negotiating its EU membership, was the possible conflict of the cap with the state aid rules of the European Union.

Nevertheless, the exemption mechanism remains. The cap was lowered for a period to a level where the tax burden exceeded 1.2 per cent of product sales value, and substantial reductions (75 per cent) were already available when the tax burden exceeded 0.8 per cent (NMR, 1994: 95). From 2006, the 0.8 per cent relates to a threshold rather than a cap and introduces the possibility only of reduced tax rates rather than complete exemption, and furthermore the reductions are limited to coal and gas. The reduced rates are approximately 15 per cent (one-sixth) of the nominal rates for industry and are close to EU minimum rates.

In the region of 50–60 energy-intensive companies are believed to benefit from the threshold for reduced CO_2 taxation (NVV, 1997: 50; NMR, 2002: 100).

In contrast to the other EU member states analysed here, a requirement for agreements or voluntary measures to reduce CO_2 emissions in order to benefit from the lower tax rates above the threshold was introduced only in 2006.

6.2.2 Finland

Finland was the first country to introduce a CO_2 tax, which came into effect from 1990. The Finnish CO_2 tax was introduced with uniform tax rates for all industrial sectors and it applies according to the carbon content of fuels. One reason often mentioned for the application of a uniform CO_2 tax rate is the prevalence of hydropower and nuclear power in Finland. Still, fossil fuels contribute more than 60 per cent to energy supply, so the CO_2 tax has more impact than in Sweden, although for industry the effective tax payments are comparable.

No special exemption mechanism for energy-intensive industries accompanied the initial Finnish CO_2 tax scheme. The reason for the absence of mitigation measures seems to have been the relatively modest

---

[4] In 1990, the cap reduced overall carbon-energy taxation on manufacturing industries by 10–15 per cent.

level of the initial CO_2 tax. From 1993, however, the CO_2 tax has gradually been increased, and has recently become comparable to Swedish levels of carbon-energy taxation for industry.

In 1998, a mechanism to relieve energy-intensive industries was introduced. This mechanism involves a threshold which allows for a substantial reduction in CO_2 taxation (85 per cent) for energy-intensive industries where the carbon-energy taxation burden exceeds 3.7 per cent of value added. This tax relief is mainly to the benefit of the pulp and paper industry, where reimbursement is made to about 10–12 companies (NMR, 2002: 64).

A pioneering element of the early carbon-energy taxation scheme in Finland was the principle of fuel-input taxation for electricity production rather than a conventional end-user tax. As the tax scheme applied a uniform tax rate for imported electricity, it was found to be in conflict with EU competition rules (see European Court of Justice, 1998). In 1997, Finland changed the tax base and an end-user electricity tax has since applied. The end-user electricity tax is unable to discriminate according to carbon emissions and so the environmental effectiveness of the energy taxation scheme has, with regard to electricity, lost its original precision.

6.2.3 Denmark

The mitigation schemes in Sweden and Finland are relatively simple and involve few companies; in contrast, the scheme which has been introduced in Denmark is comprehensive and complex. Whereas the CO_2 taxes in Sweden and Finland altered the tax base of existing industrial energy taxes, no taxation of industrial energy consumption was in place in Denmark prior to the introduction of the CO_2 tax in 1992. The novelty of the introduction of taxation of industrial energy consumption and the accompanying political negotiations may help explain why competitiveness concerns were more outspoken in Denmark and why mitigation measures became more prevalent.

For industry, the standard Danish CO_2 tax rate of DKK 100 (EUR 13) per tonne for fuels is less than half the reduced Swedish tax rate for CO_2 from fuels. In the first phase in Denmark, from 1992–5, a further 50 per cent refund of the CO_2 tax was in fact possible for all sectors. In addition, tax payments exceeding a threshold of 3 per cent of net sales value were reduced by 90 per cent. As a result, the effective CO_2 tax rate for heavy industry was 5 per cent of the nominal rate.

In the second phase, from 1996, mitigation measures were constrained. The new system introduced three different tax rates for standard industrial processes, heavy industry processes, and heating purposes, respectively. The standard rate continued at DKK 100 per tonne CO_2 and reductions were phased out, except for industries that committed themselves to energy efficiency measures by means of binding agreements. However, with agreements in place, heavy industry continued to benefit from arrangements that allowed them to lower their tax rate to 3 per cent of the standard rate, which must be regarded as a rather favourable reduction even compared with the initial system. It is estimated that about 100 companies benefit from reduced rates for heavy industries and that these companies are responsible for more than half of the industrial sector's emission of CO_2 in Denmark.

Only the tax rate for industrial space heating is similar to tax rates paid by households (about DKK 600 (EUR 80) per tonne CO_2). An end-user CO_2 tax on electricity reflects the average CO_2 content of the Danish energy supply system and applies equally to households and industry, with a rate of EUR 0.012 per kWh. This is similar to the level applied in the Dutch approach, but higher than in the other Nordic countries, reflecting the higher carbon content in Danish electricity generation.

A particular feature of the Danish mitigation approach has been the recycling of revenues for energy efficiency purposes. Twenty per cent of the revenue generated by the CO_2 tax on industries has been recycled in energy efficiency measures (while the remaining revenue has been used mainly to reduce employers' social security contributions). Annually, from 1996–2001, more than 1,000 industries received energy efficiency subsidies on the basis of CO_2 tax revenues. These generally required co-financing and an internal rate of return allowing for a four-year depreciation period.

6.2.4 Netherlands

The Netherlands first hesitated to introduce an environmental tax reform, and for several years the regulatory approach applied to energy-intensive companies was one of binding long-term sectoral agreements between government and energy-intensive sectors.

In 1996, an energy tax focusing mainly on small-scale consumers was implemented and instituted a carbon-energy tax scheme along the lines of the European Commission's proposal for EU member states in general. This tax on small-scale consumers combined with two existing taxes, increasing energy taxation so that it constitutes an important source of revenue. The tax on small-scale users applies to a range of energy products, but tax rates for two important energy products have been mitigated: gas and electricity. The tax scheme provides a cap on taxes above certain consumption thresholds (initially 170,000 m^3 natural gas or 50,000 kWh electricity), implying lenient treatment of energy-intensive industries. These thresholds have been adjusted several times and are now 1 million m^3 for gas and 10 million kWh for electricity.

Most companies are affected by the tax on small-scale energy consumption, but despite the higher thresholds, approximately 60 per cent of industrial energy consumption of gas and electricity benefits from the reduced rates.

From 2001, a zero rate above the thresholds is no longer available, but tax rates are reduced according to a scheme that differentiates among different consumption level groups. From 2004 and following the implementation of the Energy Taxation Directive, energy-intensive industries are effectively liable to the European minimum rates for energy products for consumption above the thresholds (European Commission, 2003).

For smaller users, Dutch tax rates are moderate compared with Nordic countries; the level for natural gas (1–10 million m^3) is about half the level in Denmark. For electricity, however, tax rates in The Netherlands are comparable to Danish rates for industry, and are significantly higher than those in Finland and Sweden.

6.2.5 Slovenia

The Slovenian CO_2 tax, which was introduced in 1997, supplemented the former *ad valorem* energy taxation of liquid fuels. It was extended in 1999 and 2000 also to include excise duties for transport fuels and natural gas. More than 50 per cent of Slovenia's electricity is produced by hydropower or nuclear units, however, and electricity has not been subject to CO_2 taxation per se. However, from 1992–9, electricity was subject to a 5 per cent non-deductible sales tax that also applied to industry (Ministry of Environment, 1997: 132) and from 2007 the EU minimum rate has been introduced.

There has been a complex range of reductions available from the CO_2 tax. Certain energy-intensive industries, including power plants with more than 10 tonnes of annual CO_2 emissions, were allowed a basic deduction according to their baseline emissions. Coal used for power generation has been explicitly exempt. In addition, CHP (combined heat and power) units received a tax reduction and reductions were also available for district heating. Specific companies producing heat insulation materials and transport installations for natural gas have been exempt. In total, around 150 companies benefit from direct reductions; however, a broader range of companies benefit indirectly from the treatment of the electricity generators. The Slovenian government has indicated in their report to the UNFCC that the value of exemptions to industry amounts to 67 per cent, and this figure is also mentioned in an independent report (Klemenc *et al.*, 2002). It is not clear whether this share refers to the share of electricity consumption or whether some fuels are included too.

The situation changed fundamentally when Slovenia joined the European Union in May 2004. The European Commission investigated the compatibility of the exemptions with the Energy Taxation Directive and state aid rules. As a result, the scheme was changed from 1 May 2005, with reductions for specific companies being phased out over the subsequent five years. Reductions for power plant fuels and CHP operators remain in place, as they are in accordance with the principles of the Energy Taxation Directive. Although the Commission was not explicit in its decision, it seems that reductions for power plants are admissible because the Energy Taxation Directive requires an end-user tax for electricity. According to the accession agreement, Slovenia introduced the minimum rates for electricity taxation of industries by 1 January 2007.

The nominal tax rates in Slovenia for liquid fuels are comparable to those found in other member states. Energy-intensive industries have apparently not received special mitigation treatment under Slovenia's CO_2 taxation scheme, except for a recent attempt to exempt ETS sectors.

6.2.6 Germany

Germany's Environmental Tax Reform, which was introduced in 1999, extended an earlier system of taxation of liquid energy products for industry and transport. Although its introduction was highly publicized, the tax reform implied merely that, for electricity, the *Kohlepfennig*, which had been phased out in 1996, was from 1999 to be succeeded by a new environmental electricity tax. The current tax base is not adjusted according to the carbon content of fuels, however, and it is of note that coal as fuel for industrial purposes was not taxed until 2006. The main revenue seems to accrue from the transport sector rather than from industry.

While the nominal tax rates per tonne of CO_2 for liquid fuels and natural gas are comparable to those of Sweden and Finland, the reduction mechanisms for energy-intensive industries result in much lower effective tax rates for companies. In the German system, there is both a cap on tax payments and a threshold for peak adjustments above which significantly reduced rates apply.

The cap During the first four years, from 1999–2002, the net tax rates for all manufacturing industries, as well as for the agricultural, fishery and forestry sectors, were set at 20 per cent of the nominal rates. From 2003, the net rates have been adjusted to 60 per cent of the nominal rates.

The peak adjustment (spitzen-ausgleich) In addition to this cap arrangement and in order to mitigate specifically the impacts on energy-intensive industries, the German system also offers a second option, the special *spitzen-ausgleich* (peak adjustment), to industries that otherwise would experience a net increase in taxation when considering tax relief from reduced social security contributions in relation to increased energy tax rates. The derogation initially guaranteed a full reimbursement above a 20 per cent net tax increase, but since 2003, reimbursement instead has been limited to 95 per cent of the full tax increase to retain a more balanced incentive for energy efficiency. The peak adjustment applies only to the energy tax increases introduced from 1999 and onwards.

The combined effects of the exemptions work out differently for different sectors. The exemption of coal is believed to have favoured the iron and steel industry, which consumes more than 80 per cent of this fuel. For the remaining industries, the reduced rates do not affect the energy taxes in place before 1999, so the net effect from 2003 is believed to be a reduction in nominal rates to approximately 60 per cent rather than 40 per cent. However, for industries that benefit from the peak adjustment arrangement the reductions are more significant, as is clear from the comparative analysis of implicit carbon-energy tax rates in Speck and Jilkova (Chapter 2, 37 ff). Nevertheless, the German authorities have shown that the reduced rates are sufficient to meet the minimum rates of the EU's Energy Taxation Directive. As with Sweden and The Netherlands, the reduced rates for energy-intensive industries align with the minimum rates of the EU's Energy Taxation Directive.

The first type of reduction (the reduced rate for all manufacturing industries, etc.) is in accordance with EU state aid regulations, but is limited to a period of ten years.

For energy-intensive industries, the second type of reduction (peak adjustment) is conditional on the fulfilment of the voluntary agreements concluded with industries which have established targets for energy efficiency. The European Commission has extended the exemptions, but has required additional targets to be established, along with a system of penalties for non-compliant parties.

The German government estimates the annual value of the peak adjustment at EUR 2 billion for German industry. The value of all the exemptions to industry come to about EUR 5.7 billion, as compared with total ETR revenue in Germany of EUR 18 billion.

6.2.7 UK

In the UK, the Climate Change Levy (CCL) was adopted in 1999 and introduced in 2001. It applies to gas, coal, electricity, and LPG for industry and commerce, while households are exempted. Liquid fuels are not covered, as they are covered by the hydrocarbon oil duty. Nominal tax rates are among the lowest in the seven countries reviewed here. The CCL succeeded the fossil fuel duty for electricity. The introduction of the CCL can be regarded as an implicit ETR in so far as the revenue it brings would otherwise have required an increase in other taxes, but there was not an explicit 'tax shift' as such.

Reductions (80 per cent) are available for energy-intensive industries as classified under the EU's IPCC Directive. The sectors comprise cement, aluminium, ceramics, chemicals, food and drink, foundries, glass, non-ferrous metals, paper, and steel, and 30 smaller sectors.

There is a requirement to comply with stringent energy efficiency agreements that are negotiated with the sector associations. The results of the agreements are reviewed, and continued discounts rely on targets being achieved.

Part of the revenue (approximately 5 per cent) is channelled to investments in energy efficiency via the Carbon Trust.

For electricity, the CCL provides an end-user tax, but there is an exemption for 'new' renewables (wind, solar, etc., but not larger hydropower plants or waste incineration). There is also an exemption for 'good quality' CHP.

6.3 Ex-post compensation: revenue recycling approach

As noted in the introductory chapter (Andersen, Chapter 1, 3 ff) it makes a significant difference whether revenues are recycled via a lowering of income taxes for wage earners, or via a lowering of employers' social security contributions. The so-called tax interaction effect (Bovenberg and de Mooij, 1994) would suggest that revenue recycling under the second method is more desirable, as inflationary impacts can be minimized.

On the basis of a more detailed review of revenue recycling methods in Europe's ETR countries (cf. Speck and Jilkova, Chapter 2, 24 ff), it is possible to make the following observations:

- Sweden and Finland have mainly recycled revenue by lowering income taxes. In Sweden, for many years, it has been an aim of tax policy to lower the pressure from income taxation on labour costs. The tax reforms in both Sweden and Finland have aimed to lower direct income taxes and carbon-energy taxes have contributed to securing alternative revenues for some, but not all, of the income tax reductions. This observation applies to Sweden's early environmental tax reform (1989) as well as the most recent phase (after 2001). It also applies to Finland in relation to the more comprehensive tax shifts introduced since 1996.

- Denmark and the UK, on the other hand, have followed the recommendations of the fiscal conventionalists more closely, for example, revenues have been aimed predominantly at lowering employers' social security contributions, so as to avoid inflationary effects. However, because of the imbalance between energy consumption on the one hand and the number of employees on the other, the lowering of social security contributions does not necessarily, at company level, lead to full compensation. The imbalance has, in Denmark as well as in the UK, been mitigated via special mechanisms for energy-intensive industries, such as agreements and reduced rates for heavy industries. The real purpose of the exemptions seems to have been to avoid tax interaction effects. Finally, both countries have earmarked some revenue (5–20 per cent) for direct energy efficiency subsidies, for example, via the Carbon Trust in the UK, perhaps out of concern that incentives would otherwise be too weak.

- The Netherlands and Germany have pursued a 'mixed' approach. The Dutch reduced income taxation in the initial phase, a particular characteristic being social concerns that led to an increase in the basic tax-free allowance for income as well as to complicated formulae for exempting basic consumption of electricity and gas (Vermeend and van der Vaart, 1998: 11). In the second phase, the Dutch stuck more closely to the fiscal conventionalists' advice and reduced employers' social security wage component, but they also reduced corporate taxes. In Germany, ecological tax reform split the revenue recycling equally between a reduction in employers' and employees' social security contributions, thereby establishing a programme of revenue recycling concerned less with fiscal orthodoxy and more with political appeal, taking into account that eco-tax reform was aimed at gasoline as much as at other fuels.

• Slovenia, according to its official report to the UNFCC (Ministry of the Environment, Spatial Planning and Energy, 2002: 4), does regard its package of increased carbon-energy taxes as a 'green tax reform', but the authors of this chapter have not been able to trace the specific revenue recycling approach.

Hence we can distinguish three different approaches to revenue recycling: the fiscal conventionalists (UK and Denmark), the fiscal pragmatists (Sweden and Finland), and finally, the political pragmatists (Netherlands and Germany). The pragmatists are so labelled, because their reforms were designed so as to accommodate the prevalent pressing concerns with the tax systems from the electorate, rather than with fiscal theory. Conversely, the fiscal conventionalists have adhered more to the viewpoints of fiscal theory.

6.4 Winners and losers in ETR

In the following, we explore the premises of the exemptions and special arrangements from a sector perspective: what are the actual mitigated costs of ETR to industries and to what extent have these costs been compensated for by revenue recycling through lowering employers' social security contributions (SSC)? From a company perspective, the increased level of carbon-energy taxation will be offset by two factors: (1) revenue recycling by reducing SSC, and (2) improved energy efficiency, which leads to lower unit energy costs (cf. the elasticities derived in Enevoldsen *et al.*, 2007). There is a third factor at play, the so-called Porter effect, that is, the increase in value added as a result of the pressure to innovate and become more competitive (cf. Andersen, Chapter 1, 3 ff). The

Table 6.1. The net costs of ETR as a percentage of gross operating surplus, taking into account revenue recycling to employers as well as the share of improved energy efficiency related to the increase in carbon-energy taxes

	Meat	Paper	Chem.	Pharm.	Glass	Cement	Ferrous	Non-ferrous
Denmark	−0.8	0.0	−0.1	−0.1	−0.3	1.4	−2.3	−0.9
Germany	6.9	1.2	−1.2	1.1	0.2	−0.4	−1.6	−2.1
Sweden	0.0	0.0	−0.5	0.0	−1.5	−3.7	−2.9	−0.3

Note: Data for Denmark are for 1996–2001; data for Germany are for 1999–2002; and data for Sweden are for 1996–2002.

existence of such an effect is suggested both by macroeconomic modelling results (Barker *et al.*, Chapter 7) as well as by panel regression analysis (Enevoldsen *et al.*, Chapter 5).

Table 6.1 provides an overview of the share of ETR net expenditures at the sectoral level as a share of gross operating surplus[5] (GOS) for three countries for which revenue recycling data could be disaggregated to the sectoral level. The revenue data obtained from national sources has been split into sectors according to Eurostat employment data.

It is evident that for most sectors ETR appears to represent a cost, even when the accelerated energy efficiency improvements which can be related to the tax increases are taken into account. However, the energy productivity improvements which can be related to the tax increases are relatively minor compared with the gross energy productivity improvements that have taken place during the last decade. Table 6.1, therefore, provides a careful estimate of the net balance of ETR, without taking into account the Porter effect as such. (The elasticities derived for energy savings in the Nordic sectors have also been applied to German sectors.)

When interpreting the results of Table 6.1, it needs to be borne in mind that Sweden did not recycle revenue via a lowering of SSC, but via lowering of income taxes. Nevertheless, the burden of ETR falls mainly on the most energy-intensive sectors, for example, glass, cement, and ferrous metals. Swedish cement, in particular, has not benefited from the same exemptions as elsewhere in EU. The impacts of revenue recycling via lowering of income taxes on salary levels cannot be accounted for here (readers interested in the broader macroeconomic view are referred to the E3ME results for Sweden, cf. Barker *et al.*, Chapter 7).

[5] Gross operating surplus denotes the surplus of activities before consumption of fixed capital.

In the case of Germany, approximately 50 per cent of the revenue was recycled via lowering employers' SSC and so the burden mainly accrues to the most energy-intensive industries, in particular ferrous and non-ferrous metals. However, the figures do not incorporate the *spitzen-ausgleich* exemption mechanism, that is, the thresholds for peak tax burdens, and so Table 6.1 actually overestimates the net costs of ETR for energy-intensive industries in Germany (the same caveat applies to Figures 6.2 and 6.5). The German ex-post compensation scheme is rather complex, and more detailed national studies (Bach, 2005) have made attempts to account for the *spitzen-ausgleich*.

In the case of Denmark the complex tax exemption mechanisms, combined with the unusually high tax rate for heating, have evened out the tax burden between sectors, but ferrous industries appear to have experienced some inroads on their gross operating surplus. Cement, surprisingly, has accomplished a positive net benefit from ETR; this is due to the substantial fuel shifts carried out in the sector (in particular substitution to the use of waste as fuel) and the energy efficiency improvements attained.

Figures 6.1, 6.2, and 6.3 decompose the net effects of ETR for the three countries into the gross carbon-energy tax payments, the revenue recycling, and the gains from improved energy efficiency, respectively.

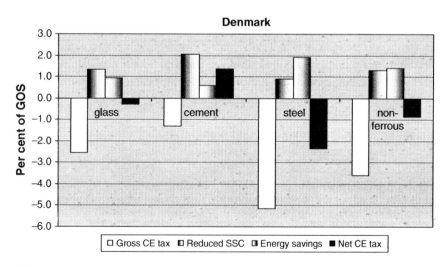

Figure 6.1. Decomposition of the net effects of ETR for Denmark into gross carbon-energy tax payments, revenue recycling, and gains from improved energy efficiency

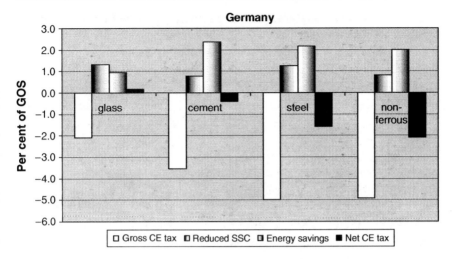

Figure 6.2. Decomposition of the net effects of ETR for Germany into gross carbon-energy tax payments, revenue recycling, and gains from improved energy efficiency

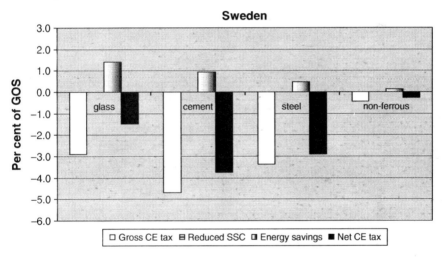

Figure 6.3. Decomposition of the net effects of ETR for Sweden into gross carbon-energy tax payments, revenue recycling, and gains from improved energy efficiency

The decomposition of the ETR costs at the sectoral level shown in Figures 6.1, 6.2, and 6.3 for Denmark, Germany, and Sweden shows that ETR, even with the exemption mechanisms in place, induces a gain for energy-intensive industries only in exceptional cases. The general pattern is one of a burden for the most energy-intensive industries. Conversely, the less energy-intensive industries (meat, pharmaceuticals, paper products) have managed to offset the costs of ETR; however, substantial gains are not apparent.

Some degree of revenue recycling from employers' SSC is an important measure to reduce the direct costs of ETR, as reflected in the lower net costs for Danish and German industries as compared to Swedish industries. Still, the importance of revenue recycling should not be overemphasized, as in six of the eight Danish and German sectors, the savings via improved energy efficiency are more significant than the revenue recycling itself. From the sectoral perspective, the burden on energy-intensive industries is negative, but moderately so. For cement and glass, it is less than 1 per cent of the gross operating surplus when there is some revenue recycling of employers' SSC, while for ferrous and non-ferrous metals it appears to have reached in some cases 2 per cent of gross operating surplus. In the Swedish case, with no SSC revenue recycling, the costs are estimated to be higher: up to 4 per cent of gross operating surplus for cement and steel. Company managers in energy-intensive industries may not have appreciated the tax-induced improvements in energy efficiency and may have focused more on the gross burden of ETR, which unadjusted for the gains, has represented up to 5 per cent of the gross operating surplus for some energy-intensive industries in all three countries.

The claim of the Porter hypothesis (Porter, 1991) is actually not that energy taxation will induce sufficient energy savings to even out the increased tax burden. Porter's hypothesis is rather that increased carbon-energy taxation will in the longer term pressure industries to innovate both in their processes and products so as to become more competitive and win market shares.

In the COMETR project, both the E3ME modelling of the macroeconomic impacts and the panel regression analysis of the impact of energy taxes in 56 industrial sectors pointed to the existence of such 'hidden' Porter demand effects. In the following, we put the sectoral costs of ETR (cf. above) in perspective in relation to these 'Porter effects', as well as the gross energy savings attained by industries in the wake of ETR.

The gross energy savings are the costs foregone per GJ of output at current energy prices. Bearing in mind that above, only the *accelerated* energy savings that could be attributed directly to annual tax rate increases were included, we show here the value of gross energy savings achieved by the various sectors. The additional energy efficiency savings attained in most sectors are far higher than can be attributed statistically to the tax rate increases. As energy prices were relatively stable over the period analysed here, changes in underlying fuel prices cannot explain the savings.

In most cases where ETR led to increased costs, these have been more than offset by gross energy savings. To some extent, the gross energy savings reflect 'business-as-usual' and only those energy savings attributed to tax rate increases should be included when accounting for the impacts of ETR, as approached above. Nevertheless, the gross energy savings achieved do put the costs of ETR in an illuminating perspective.

Figures 6.4, 6.5, and 6.6 provide an overview of the costs of the ETR burden relative to the gross energy efficiency savings. In addition, the three figures provide an estimate of the Porter demand effect on the basis of the relationships derived in the panel regression analysis, which identified a statistically significant relationship. However, as a minor degree of multi-collinearity in that analysis could not be ruled out, the Porter demand effects must remain a best guess and their quantification would require further efforts with improved econometric techniques.

First of all, the costs of ETR—now as a share of gross value added (GVA)—are, in practically all sectors, an order of magnitude lower than the gross energy efficiency savings attained, as well as the estimated Porter

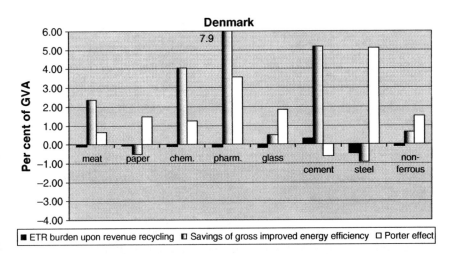

Figure 6.4. Overview of the costs of the ETR burden (for Denmark) relative to gross energy efficiency savings and a possible Porter effect

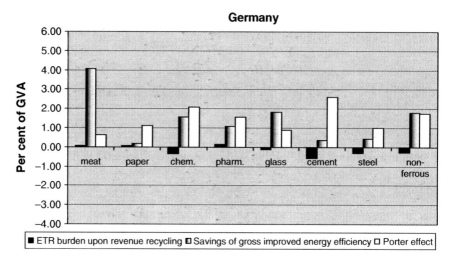

Figure 6.5. Overview of the costs of the ETR burden (for Germany) relative to gross energy efficiency savings and a possible Porter effect

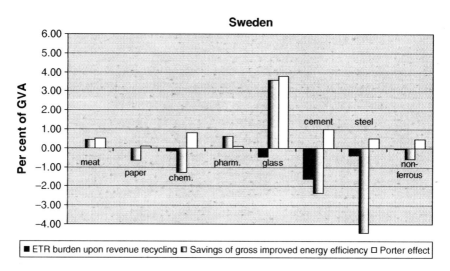

Figure 6.6. Overview of the costs of the ETR burden (for Sweden) relative to gross energy efficiency savings and a possible Porter effect

effects. Important exceptions to this general trend can be observed for the cement and steel industries. Here the gross energy savings are not impressive in relation to the ETR burden, apart from in the case of Danish cement.

As noted above, the ETR costs for Germany are overestimated, as the value of the *Spitzen-ausgleich* has not been included. As the ETR costs as a share of GVA are nevertheless very modest, this observation is without implications for the following inspection of the differences between Denmark, Germany, and Sweden, which in some ways are striking and deserve attention.

For chemicals, pharmaceuticals, and cement, gross energy savings are far more significant in Denmark than in either of the other two countries. Conversely, Germany leads with regard to energy savings in ferrous and non-ferrous metals and meat. Sweden excels in its glass industry only, while several other sectors saw their energy efficiency deteriorate.

Swedish steel is an interesting case, as energy consumption has increased, particularly that involving the use of coal and coke, while economic output has remained constant. It seems that fuel switches introduced during the initial, 'idealized' carbon-energy taxation scheme applied in 1992–3 were reversed in the latter half of the 1990s, which may help explain the deterioration in performance of this and other Swedish sectors (as well as the exemption mechanism for coal and coke in ferrous and non-ferrous metals).

For Germany, ETR was initiated as late as 1999 and has been in operation for a shorter period of time than ETR in Denmark and Sweden, analysed here for the period from 1996 (and for all three countries up to 2002). Previous research has shown that the time span required for adaptation to increased energy taxation is approximately four years, so the time span should be sufficient to capture the full effects in Germany. In Germany, 50 per cent of revenue has been recycled to lower employers' social security contributions, and this helps to create a positive ETR balance for three sectors, even without considering improved energy efficiency savings and Porter effects.

To sum up, while for most energy-intensive sectors the tax-induced energy savings were not sufficient to offset the ETR burden, this burden remains for most sectors an order of magnitude lower than the overall energy savings accomplished during the years of ETR.

The main problem appears to be with the cement and steel sectors, which seem to have had some difficulties absorbing the ETR burden; although Danish cement stands out as a notable exception to this pattern, with its considerable energy efficiency savings. These savings were achieved by lowering energy intensity from approximately 67 GJ per EUR 1,000 output to a level of 50 GJ per EUR 1,000 output in just seven years, that is by 25 per cent. Still Sweden's energy intensity for cement is at about the same level as Denmark's, and Germany's is even lower (40 GJ per EUR 1,000 output), so the pattern for Danish cement may reflect that a backlog of improvements was drawn upon.

Swedish and German cement have been subject to higher tax burdens than Danish cement (EUR 0.35 per GJ and EUR 0.21 per GJ, respectively, versus EUR 0.05 per GJ in Denmark (Ryelund, 2007)), nevertheless cement's energy efficiency has not improved markedly in the two former countries. It seems that Swedish cement was able to absorb the tax through a lowering of its energy costs by switching fuels, hence keeping overall energy costs roughly constant. In Denmark, both fuel switching and energy savings were involved. The findings lead to the suggestion that more substantial tax rates would be required to induce further energy savings, and that the industry might be facing a technology threshold that would require additional efforts to transcend. A recent IEA report (2007) states that by switching to dry process rotary kilns from traditional wet process technologies, the energy efficiency of cement industries could be improved by up to 50 per cent. Investments required for cement plants are significant and would amount to approximately three years of turnover (Jilkova *et al.*, 2007). As cement accounts for about 10 per cent of total final energy use in industry, the potential contribution to energy savings from the use of best available technology is by no means trivial. Cement is not unequivocally a price-taker; the value-to-weight ratio of cement does not allow for long-distance land transport. Direct access to port facilities can extend the range of trade activities, however. After the food processing sector, the non-metallic minerals products sector, the parent sector of cement, is the least trade intensive (cf. Fitz Gerald *et al.*, Chapter 3). This sector is also ranked as the least sensitive in terms of price-setting power.

With respect to steel, it is second to cement in energy intensity, with levels varying from 19 GJ per EUR 1,000 output in Germany over 13 GJ per EUR 1,000 output in Sweden and only 5 GJ per EUR 1,000 output in Denmark, where plants rely mainly on the technology of electric arc furnaces. The differences are believed also to reflect differences in average plant size and as well as product characteristics (including wider use of scrap steel in Denmark). In terms of economic output, the industries are not declining and they cannot be termed 'sunset' industries, as such. For steel, we identify the highest effective tax burden in Denmark (0.77 EUR per GJ), as opposed to approximatley 0.27 EUR per GJ in Sweden and Germany. As mentioned above, energy efficiency deteriorated in Sweden's steel industry. Moreover, the same trend, although less pronounced, has been identified for Denmark, whereas for Germany a very moderate increase can be identified for the steel sector. The effective tax burden per GJ for the steel sector is at the same level as tax burdens in other sectors, including the energy-intensive chemicals sector. A closer inspection of the energy costs suggests that the modest tax burdens have been absorbed by fuel shifts that entailed a lowering of energy expenses. As this decrease more or less offsets the increased tax burden, no net improvement in energy efficiency and productivity is evident.

A recent IEA report which reviews technology options in the steel industry shows that a broad range of technological processes are employed in the sector (IEA, 2007: 108). The traditional basic oxygen furnace (BOF) method is one of the most energy and carbon intensive, and the options for improvement in energy efficiency are relatively limited. A switch to use of electric arc furnaces based on gas would entail more significant savings in relation to carbon emissions. Furthermore, by switching from pig iron to use of scrap iron in traditional electric arc furnaces, CO_2 emissions per tonne of steel can be reduced to 20 per cent of the level with the traditional BOF method. The main issue here is that the method is subject to the constraint of the limited availability of scrap iron of suitable quality.

Conventional BOF methods continue to account for two-thirds of production capacity in Europe, while, for example, electric arc furnaces account for only approximately 30 per cent of steel production in Germany and 20 per cent in the UK. As electric arc furnaces rely on electricity rather than coal, the production method can be based on hydropower and gas, as is the case in Slovenia and Denmark, rather than coal, as predominantly used in the sector in Germany and Sweden (Christie *et al.*, 2007: 31). As the iron and steel industry is clearly a price-taker, limited opportunities exist to pass on the costs of carbon, if factored into the cost structure via ETR. However, a more phased introduction of ETR, with some revenue recycling for an investment programme to renew production technologies, would allow for an implicit fuel shift in favour of electric arc furnaces. Improved levels of steel recycling would furthermore increase the capacity of scrap-based steel processing and, as the sector's location decisions are tied more to the availability of iron than to energy requirements, this might support the sector in continuing its production activities within the EU.

6.5 Conclusions

The special arrangements that have developed within the unilaterally introduced ETRs have, unfortunately, replaced the transparency and calculation methods of economic incentives with a rather thick fog of exemptions of the kind that materializes when the heat of vested interests meets the mists of tax legislation.

In terms of ex-ante measures for mitigation, all countries have offered energy-intensive industries exemptions and reductions in relation to the tax base and tax rates, but the specific exemptions vary to some extent between member states. Sweden, Finland, and The Netherlands have a threshold in their carbon-energy tax legislation, above which the national addition to the EU minimum rates does not apply to large energy users. However, while in the Dutch case, the threshold is in place generally for all large consumers of gas and electricity, in Finland and Sweden, the minimum rates apply only to industries where the tax burden exceeds a predefined share of gross value added.

In Germany, the UK, and Denmark, industries with energy efficiency agreements are entitled to refunds on their tax payments; however, while in Germany the minimum rate remains at 60 per cent of the nominal rate, the payments in the UK are generally at 20 per cent, while in Denmark they can be as low as 3 per cent. Also, the exemptions for specific fuels differ considerably; while in Germany the use of coal remains tax exempt for households and certain industrial uses, this fuel is subject to taxation in most other countries—most of which grant benefits under the Energy Taxation Directive (2003/96/EC) exemptions for metallurgical industries and power plants. The general and specific energy tax rules combine to produce a complex mosaic of exemptions, making it difficult to make generalizations about the prevailing conditions—conditions which vary considerably between sectors, energy carriers, and countries.

As to the ex-post approaches for compensation, the countries can be divided into three groups: the fiscal conventionalists (Denmark and UK), which have adhered to revenue neutrality through a lowering of employers' social security contributions; the political pragmatists (Sweden and Finland), which have preferred to lower their effective income tax rates; and finally the fiscal pragmatists (Germany and The Netherlands), which have chosen to combine lowering of social security contributions (both employers' and employees') with lowering income taxes.

The European Union's Energy Taxation Directive, as well as the European Commission's Guidelines for State Aid (2008), have in recent years (since 2003) been helpful in narrowing the differences in mitigation, but have had little harmonizing impact on ex-post compensation. In particular, the decisions on how to recycle revenues are entirely within the discretion of member state competencies. In this chapter, we have explored the effective sectoral burdens of ETR for energy-intensive industries with respect to three countries: Denmark, Germany, and Sweden (one from each of the three above-mentioned categories of ETR). Without taking either revenue recycling or energy efficiency into account, the burden of ETR for energy-intensive sectors, net of the value of exemptions and reductions, has not exceeded 5 per cent of gross operating surplus in any sector in these countries. For Denmark and Germany, the net burden, taking into account the value of the revenue recycling of employers' social security contributions and tax-induced energy efficiency measures, has not exceeded 2 per cent of gross operating surplus for the most negatively affected sectors, ferrous and non-ferrous metals. For other energy-intensive industries, glass and cement, the burden has been in the region of 1 per cent. These figures do not include the German peak-tax adjustment and so represent a conservative estimate of the costs for Germany.

Overall, the net costs of ETR have, in most sectors, been exceeded by the value of the gains in energy efficiency which have been obtained over the same period of time. The exceptions to this pattern are Danish steel, the German steel and cement sectors, as well as several Swedish energy-intensive sectors, where energy efficiency improvements have not been sufficient to offset the burden imposed by ETR. The troubled history of ETR in Sweden is believed to have produced a backlash, as energy-intensive industries increased energy consumption in response to the marked reduction in CO_2 taxation in relation to the initial level in 1991–2. The Swedish steel sector, for instance, increased its energy consumption following the reduction of CO_2 taxation from 1993, while economic output remained constant. In Denmark and Germany, on the other hand, the costs of ETR have been offset by gains in energy efficiency, while the potential Porter effect (improved competitiveness) has added to these gains.

References

Bach, S. 2005., Be- und Entlastungswirkungen der Ökologischen Steuerreform nach Produktionsbereichen'. Berlin: Deutsches Institut für Wirtschaftsforschung.

Bovenberg, A. L., and de Mooij, R. A. 1994. 'Environmental levies and distortionary taxation'. *American Economic Review*, 84: 1085–9.

Christie, E., Hanzl, D., and Scott, S. 2007. 'Case study on the iron and steel industry', in M. S. Andersen, T. Barker, E. Christie, P. Ekins, J. Fitz Gerald, J. Jilkova, J. Junankar, M. Landesmann, H. Pollitt, R. Salmons, S. Scott, and S. Speck, *Competitiveness Effects of Environmental Tax Reforms (COMETR): Annex to Final Report to the European Commission*, DG Research and DG TAXUD. National Environmental Research Institute, Aarhus University.

Enevoldsen, M. K., Ryelund, A. V., and Andersen, M. S. 2007. 'Decoupling of industrial energy consumption and CO_2-emissions in energy-intensive industries in Scandinavia'. *Energy Economics*, 29/4: 665–92.

European Commission (EC). 2003. Council Directive 2003/96/EC of 27 October 2003 restructuring the Community framework for the taxation of energy products and electricity. OJ L283, 31.10.2003, pp. 51–70.

—— 2008. Community Guidelines on State Aid for Environmental Protection. OJ C82, 1.4.2008, pp. 1–33.

European Court of Justice. 1998. Judgment (*Outokompu Oy* case). C-213/96, Luxembourg.

International Energy Agency (IEA). 2007. *Tracking Industrial Energy Efficiency and CO_2 Emissions*. Paris: IEA.

Jilkova, J., Pisa, V., and Christie, E. 2007. 'Case study on the cement, lime and plaster industry', in M. S. Andersen, T. Barker, E. Christie, P. Ekins, J. Fitz Gerald, J. Jilkova, J. Junankar, M. Landesmann, H. Pollitt, R. Salmons, S. Scott, and S. Speck, *Competitiveness Effects of Environmental Tax Reforms (COMETR): Annex to Final Report to the European Commission*, DG Research and DG TAXUD. National Environmental Research Institute, Aarhus University.

Klemenc, A., Merse, S., and Tomsic, M. 2002. 'Slovenia: the carbon-dioxide tax and investment in co-generation plants', in E. Petkova and D. Faraday, *Good Practices in Policies and Measures for Climate Change Mitigation*. Budapest: Regional Environment Centre (REC) and World Resources Institute, 89–99.

Ministry of Environment. 1997. 'Environmental performance review for Slovenia'. Ljubljana.

Ministry of the Environment, Spatial Planning and Energy. 2002. 'Slovenia's first national communication under the UN framework convention on climate change. Ljubljana.

Naturvårdsverket (NVV). 1997. Miljöskatter i Sverige. Stockholm.

Nordic Council of Ministers (NMR). 1994. 'The use of economic instruments in Nordic environmental policy'. Copenhagen.

—— 2002. 'The use of economic instruments in Nordic environmental policy 1999–2001'. Copenhagen.

Organisation for Economic Co-operation and Development (OECD). 2001. *Environmentally Related Taxes in OECD Countries: Issues and Strategies*. Paris: OECD.

Porter, M. 1991. 'America's green strategy'. *Scientific American*, 264: 168.

Ryelund, A. 2007. 'Improvements in energy efficiency and gross carbon-energy tax burdens in eight energy-intensive and less energy-intensive sectors: a sub-sector perspective', in M. S. Andersen, T. Barker, E. Christie, P. Ekins, J. Fitz Gerald, J. Jilkova, J. Junankar, M. Landesmann, H. Pollitt, R. Salmons, S. Scott, and S. Speck, *Competitiveness Effects of Environmental Tax Reforms (COMETR): Annex to Final Report to the European Commission*, DG Research and DG TAXUD. National Environmental Research Institute, Aarhus University.

SOU. 1991. Konkurrensneutral energibeskattning, 1991/90. Stockholm.

Vermeend, W., and van der Vaart, J. 1998. *Greening Taxes: The Dutch Model*. Deventer: Kluwer.

Part III

Country Competitiveness and Carbon Leakage

7

The Effects of Environmental Tax Reform on International Competitiveness in the European Union: Modelling with E3ME

Terry Barker,[1] Sudhir Junankar,[2] Hector Pollitt,[3] and Philip Summerton[4]

7.1 Introduction

The objective of the research discussed in this chapter is to evaluate the short-term and long-term economic effects of carbon-energy taxes introduced under environmental tax reforms (ETRs) in the macroeconomic framework provided by Cambridge Econometrics' (CE) Energy-Environment-Economy model for Europe, E3ME. The analysis undertaken in this chapter focuses upon the competitiveness effects in the energy-intensive COMETR sectors[5] for the six countries that undertook environmental tax reform in the 1990s: Denmark, Finland, Germany, The Netherlands, Sweden, and the UK. (For Slovenia, the CO_2 tax, although not strictly part of an ETR, has been included in the baseline scenario to

[1] Terry Barker, Director, Cambridge Centre for Climate Change Mitigation Research/Cambridge Econometrics, United Kingdom.
[2] Sudhir Junankar, Associate Director, Energy Environment Service, Cambridge Econometrics, United Kingdom.
[3] Hector Pollitt, Project Manager—International Modelling, Cambridge Econometrics, United Kingdom.
[4] Philip Summerton, Senior Economist, Energy Environment Service, Cambridge Econometrics, United Kingdom.
[5] The five parent NACE-2 E3ME sectors were: 15: food and food products; 21: pulp, paper, and paper products; 24: chemical and chemical products; 26: non-metallic mineral products; and 27: basic metals.

give an example of environmental taxation in the new member states.) The modelling sought to capture the inter-industry and other indirect effects, as well as international competitiveness effects, which cannot be well accounted using the bottom-up approaches. The ultimate goal is to compare the short- and long-term effects of ETR, both in terms of energy use and emissions and in terms of economic activity (see Kohlhaas, 2005, for a study of the ETR in Germany).

The main emphasis of the work has been on price and non-price competitiveness issues and the dynamics of external trade, employment, output, and investment in capital that were examined in the various ETR scenarios (see Boltho, 1996, Adams, 1997, and Barker and Köhler, 1998, for general discussions of competitiveness in international trade and Grubb *et al.*, 2002a and 2002b, for a discussion of non-price competitiveness effects). These results feed into the analysis of carbon leakage discussed in Chapter 8. The analysis has been carried out for each member state considered in the bottom-up case studies and for all the non-ETR EU countries together. The change in the sectors' costs due to a change of input composition leads to a different position for the sectors in international markets. So a particular emphasis of the work has been on the modelling of the effects of the green-tax reform on the external position of the sectors and any implication this has for the single market and EU enlargement.

The results of the modelling that we have undertaken serve two purposes:

- to identify the key characteristics of the green tax reform packages, compared with a 'reference case' (i.e. a counterfactual case) generated by E3ME over the period 1994–2012 without ETR, but including current and expected developments in the EU economy, for example, the impact of the EU ETS. The wider competitiveness effects of the reform on other sectors and in other countries (through international trade) are also assessed through the comparison of the effects from 1994 to 2012;
- to give signals to policy-makers about the relative effectiveness of different policy instruments (e.g. the full tax or levy, inclusion of exemptions and/or special treatment for affected sectors, and including and excluding revenue recycling) in overcoming the short-term costs of policies with possible beneficial long-term effects on competitiveness, as suggested by the Porter hypothesis (Porter, 1991; Porter and van der Linde, 1995).

From the long-run perspective, it is necessary to model the effects of the reform on the dynamics of technological change and investment. These issues are addressed in the context of what the analysis undertaken suggests in relation to the 'Porter hypothesis'. Porter's hypothesis is that environmental policy (especially green tax reform) can increase the international competitiveness of domestic industries in the long run, since firms are forced to adopt new, energy-saving technologies as a response to increases in energy prices (further evidence for non-price competitiveness effects is given in Reinaud, 2005; Sijm et al., 2004). The underlying assumption is that the new environmentally friendly technologies lead to a decrease in unit costs in the long run. However even if Porter's hypothesis holds, there may be significant short-term transition costs, which the policy-makers may be able to reduce, for example, by using tax refunds or supporting 'green' R&D policies.

7.1.1 Features of ETR

The environmental tax programmes differ across the six EU member states that implemented ETRs in the 1990s, in terms of the industries targeted and the revenue recycling mechanisms. The green tax reform not only changes tax rates but, because of changes in inputs, it changes also the tax base.

Green tax reform can affect one or more of the main energy-using sectors: power generation, industry, households, and transport. The taxes that increase under an ETR usually include energy taxes and other environmental taxes. Some ETRs may also involve the creation of a new tax that replaces an old one (which may not necessarily have the same tax base).

The purpose of an ETR is to shift taxation away from beneficial activities, such as employment, towards damaging activities, such as pollution. The idea is to implement specific taxes to encourage households and industries to behave in a way that is environmentally sustainable. The revenues thereby generated are used to reduce burdensome taxes to complete the ETR.

This 'recycling mechanism' may take effect through:

• direct taxes (income tax, corporation tax);
• social security contributions paid by employers;

- social security contributions paid by employees;
- other measures;
- support schemes for investment expenditure (and depreciation) and
- benefits or other compensatory measures.

In certain European countries, the ETR has also included tax provisions tailored towards certain industry sectors (particularly those that are energy-intensive) to induce a more energy-efficient consumption profile and thus reduce the environmental impacts of their economic activities.

An ETR can, in principle, provide complete tax exemption for economic sectors or reduced tax rates for different energy fuels and economic sectors, in combination with some form of negotiated agreements with targets to improve energy efficiency or carbon emissions. Tax ceilings may also be established to limit the total tax burden faced by individual companies.

7.1.2 Outline of the chapter

Section 7.2 discusses the modelling undertaken with CE's Energy-Environment-Economy Model for Europe, E3ME version 4.1, and focuses on the modelling of non-price and price competitiveness. Section 7.3 outlines the tax input data and data processing required by E3ME, while Section 7.4 describes the ETR scenarios that were specified to examine the competitiveness effects in the COMETR sectors for the six countries that undertook environmental tax reform in the 1990s. Section 7.5 reviews the results of the estimation routines in E3ME relevant to measuring the direct (and indirect) price and non-price competitiveness effects of ETRs, and also outlines the econometric theory underlying the analysis. The modelling results are presented in Section 7.6, with a detailed analysis of the key features of the macroeconomic and environmental projections in the ETR baseline scenario and also of the implications for selected industries, at the NACE two-digit level, which are analysed at NACE three-digit level elsewhere in the study. We also draw out the key findings of the analysis and consider what light is cast on the validity of the Porter hypothesis and the implications, within the E3ME framework, for the analysis of carbon leakage due to ETRs that forms the basis of the research discussed in Chapter 8.

7.2 Modelling the EU Energy-Environment-Economy System with E3ME

7.2.1 Introduction to E3ME

7.2.1.1 BACKGROUND

E3ME (Energy-Environment-Economy Model for Europe) is a general model for Europe designed to address issues that link developments and policies in the areas of energy, the environment, and the economy (see Cambridge Econometrics, 2005). The European economy is becoming more and more integrated; at the same time, the relationship between economic activity, energy use, and the environment is of pressing concern for European policy and political debate.

The guiding principles of the model are such that it is:

- elaborated at a European, rather than at a national, level, with the national economies being treated as regions of Europe;
- dealing with energy, the environment, population, and the economy in one modelling framework and allowing short-term deviations to occur while convergence to a long-run outcome takes place;
- designed from the outset to address issues of central importance for economic, energy, and environmental policy at the European level;
- capable of providing highly disaggregated short- and long-term economic and industrial forecasts for business and government;
- capable of analysing long-term structural change in energy demand and supply and in the economy;
- focused on the contribution of research and development, and associated technological innovation, on the dynamics of growth and change.

7.2.1.2 ANTECEDENTS

E3ME is a multisectoral dynamic regional econometric model capable of providing a long-term equilibrium solution. The model uses input-output tables, but combines them with the time-series analysis used in macroeconometric models. It has been developed following the structure of a regionalized E3 model of the UK economy (Barker and Peterson, 1987) which has been used to analyse in detail the effect of the EC carbon/energy tax on the UK economy. Models in the same tradition for national economies have been developed by the INFORUM group of modellers (see Almon, 1991).

151

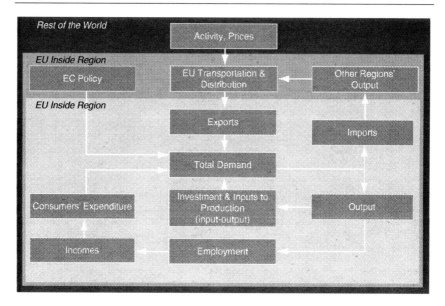

Figure 7.1. E3ME41 as a regional econometric input-output model

7.2.1.3 THE E3ME MODEL

E3ME version 4.1 is comprehensively described in the model manual (<http://www.camecon-e3memanual.com/>), which includes a full set of results from the estimated equations.

Figure 7.1 shows how E3ME can be represented as a regional, econometric input-output model. Most of the economic variables shown are at a 42-industry level (NACE two-digit with expanded fuel and power sectors, including 16 services sectors) and cover the time period 1970–2002. The whole system is solved simultaneously across all industries and countries (the EU25 in 2006 plus Norway and Switzerland). More information on the E3ME model can be found at the main model website, <http://www.e3me.com/>.

7.2.1.4 E3 MODELLING IN E3ME

Figure 7.2 shows how the three components of the model—energy, environment, and economy—fit together. Each component is shown in its own box and utilizes its own units of account and sources of data. Each data set has been constructed by statistical offices to conform to accounting conventions. Exogenous factors coming from outside the modelling framework are shown as inputs into each component on the outside edge of the chart. For the EU economy, these factors are

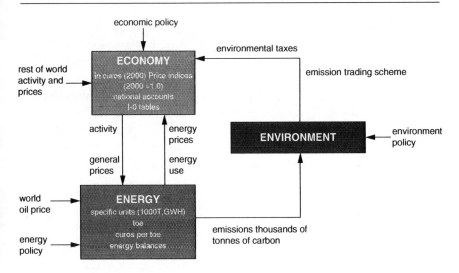

Figure 7.2. E3ME as an E3 model

economic activity and prices in non-EU world areas and economic policy (including tax rates, growth in government expenditures, interest rates, and exchange rates). For the energy system, the outside factors are the world oil prices and energy policy (including regulation of energy industries). For the environment component, exogenous factors include policies such as reduction in SO_2 emissions from large combustion plants. The linkages between the components of the model are shown explicitly with arrows showing which values are transmitted between components.

The energy price data in E3ME come from the International Energy Agency (IEA) and do not include sector-specific pricing (i.e. all industry groups are assumed to pay the same price for their various fuel inputs). Although sector-specific prices are available for some countries, they are not available on a consistent basis across Europe. However, this is a key assumption, because the expected sectoral effects of ETR will be directly related to the fuel prices paid by that sector—for example, a tax of 10€/toe will have a much larger relative effect when the fuel price excluding taxes is 50€/toe than when the fuel price is 100€/toe. This assumption means that results for sectors made up of large firms that can purchase fuels at lower prices (such as paper and pulp) are probably being understated, and those with many small firms may be being overstated.

7.2.1.5 THE DATA

E3ME's historical database is constructed using data from Eurostat, DG ECFIN (AMECO), and the OECD secretariat's STAN database. These sources have the advantage of covering the whole of the EU on a consistent basis (as far as possible), but the data are usually published later than the corresponding data from national sources. These data have been supplemented by standardized data from other sources when required, to form completed time series. For more information about the coverage of the data, the reader should refer to the model manual.

7.2.2 Modelling foreign trade in E3ME41

7.2.2.1 INTRODUCTION

The role of trade is central to analysing competitiveness in an individual country or industry. Trade is an important feature in a regional model such as E3ME for two main reasons. First, European integration has led to a rise in degree of openness in many EU markets, with an associated rise in the ratio of exports to total final demand. Second, exports and imports represent the linkage between the countries in E3ME, so any effects moving from one country to another are transmitted through this part of the model. The modelling of exports and imports is similar in structure, in terms of income and price effects and also because they use similarly constructed variables.

All trade is treated as if it takes place through a European pool. The export and import volume equations represent each country's exports into this pool and imports from it. Total exports and imports have been separated into two sub-components, one for intra-EU trade and one for extra-EU trade. However, it is not possible to identify separate trade prices for intra- and extra-EU trade, and therefore the export and import price specifications are for all exports and imports, regardless of destination or source.

The determinants of export volumes in the model can be separated into two groups of effects, those associated with income and those associated with prices. A proxy for technical progress (accumulated gross fixed investment plus R&D expenditure) is also included in the equations (see below). The basic model of trade prices used in E3ME assumes that the EU countries operate in oligopolistic markets and that each is a small economy in relation to the total market. This assumption about market structure implies that, apart from a few commodities (e.g. crude oil) whose prices are set exogenously, prices are typically set by producers as mark-ups on costs, that is, unit costs of production.

7.2.2.2 EXPORT AND IMPORT VOLUMES

Both export and import volumes are split into intra- and extra-EU trade for each country and sector, and a separate equation is estimated in each case.

The export volume equation can be separated into three effects, income, prices, and technical progress. The income effect is captured in the form of two variables, the first dealing with economic activity in the rest of the EU, the other concerning activity in the rest of the world. Price effects are split into three forces: the price of exports, the price of exports in other EU countries, and a 'rest of the world' price variable. All prices are converted to EUR. Homogeneity is imposed between the price effects, such that the combined value of the external price coefficients (other EU and rest of world) are set equal to the overall export price. This is another way of combining the price terms in a relative, rather than absolute, form. The technical progress indicators (see below) are also included to help capture the role of innovations in trade performance. This variable could be measured relative to that of competitors, but since this would imply no effect on trade if competitors undertook equal proportions of investment/R&D, this did not seem to be a worthwhile exercise.

In the equation describing intra-EU and extra-EU import volume, activity is modelled by sales to the domestic market, while the three price effects are import price, price of sales to the domestic market, and the relative price of the currency, that is, the EUR exchange rate. Aside from the restrictions on sign and significance, price homogeneity is imposed between the price of imports and price of sales to the domestic market. As with the export equation, this has the effect of making the price relative, removing the long-term effect of the exchange rate variable. The technical progress measure is again included to allow for the effects of innovation on trade performance.

An additional variable, SVIM, has been added to both the export and import equations to take account of the Internal Market programme. SVIM is a synthetic variable that has a value of zero until 1985, and then gradually increases (following an exponential pattern) to a value of unity in 1992 and increases further with the introduction of the EUR. SVIM is set to zero in the extra-EU trade equations.

155

7.2.2.3 EXPORT AND IMPORT PRICES

The equations for export and import prices play a large role in the response to exchange rate movements, acting as an important transmission mechanism for effects such as devaluation, for example the exit of sterling and the lira from the ERM in September 1992. The effects can be dissipated in a number of ways, for example creating inflationary pressures or leading to movements in the balance of payments. The basic model of trade prices used in E3ME assumes that the EU countries operate in oligopolistic markets and that each is a small economy in relation to the total market.

Certain commodities (e.g. crude mineral oil) have prices treated exogenously, but the majority are treated in the following manner. Following from the assumption on market structure, prices are set by producers as mark-ups on costs, that is, unit costs of production. Aside from this, the same variables are used for both import and export prices, within a general log-log functional form.

Alongside the unit cost variable, there are four price terms included in each regression to deal with developments outside the country in question. They are an 'other EU' price (created in the same manner as described in the trade volume equations), a 'rest of world' (i.e. outside EU) price, a world commodity price variable, and the EUR exchange rate itself. The measures of technical progress (described in detail below) are also included to cope with the quality effect on prices caused by increased levels of investment and R&D. Restrictions are imposed to force price homogeneity and exchange rate symmetry on the long-term equations, again in much the same manner as for the trade volume equations.

Due to the complexities and non-economic factors involved, exchange rates are treated as exogenous in E3ME. Thus a large shift in trade balances will not automatically affect the competitiveness of domestic industry. This is worth bearing in mind when considering the impact of ETR on trade prices; and gains or losses could easily be exceeded by a relatively small exchange rate movement.

7.2.3 Non-price competitiveness in E3ME41

7.2.3.1 THE ROLE OF R&D EXPENDITURE AND INVESTMENT

The technical progress indicators are used as a measure of innovation and product quality and estimate the non-price competitiveness of an industry. Ideally, E3ME could incorporate measures of innovatory activity in each EU member state, relative to the same activity in its main competitors, at a detailed industrially disaggregated level on an annual basis, covering the period 1970–2002, but limited data restrict the choice of inputs. The decision as to which data to use as a representation for innovation comes down to a choice between two alternatives: patents; and research and development activity. In each case, the available information has to be mapped onto the industrial classifications used in the E3ME model. This requires a very detailed examination and comparison of the systems of classification used in each case, sometimes involving comparisons across countries.

7.2.3.2 DATA ON PATENTS

Patenting activity represents one measure of innovatory activity. It is well established that different industries have different propensities to patent. A potentially valuable data set exists in the form of the series collected in the United States by the Department of Commerce, Office of Technology Assessment and Forecasts (OTAF). These data indicate the level of patenting activity by industry conducted by most major states within the US. Given the latter's key role in innovation and the world economy generally, these data provide a potentially very useful measure of relative innovatory activity in different countries. However, the data suffer from some limitations as far as the present exercise is concerned: the industrial classification used is a US one and, by their nature, patents tend to focus attention on manufacturing industries rather than the service sectors. It should be noted that there is ongoing work at the OECD to compile an industry database of patents, which may be a valuable input to E3ME in the future.

7.2.3.3 R&D EXPENDITURES

An alternative measure of innovation is R&D expenditure (and related employment). In contrast to the patents indicator, this is a measure of input rather than output from innovatory activity. The OECD publishes a series of data on R&D activity by industries for major economies, known officially as the ANBERD database. These, in principle, enable relative measures of member states' performance to be constructed. In practice, the OECD data are based on irregular surveys conducted within each individual country. There are therefore large numbers of missing observations. A considerable amount of interpolation and adjustment is therefore necessary to convert these data into a usable form for time-series analysis. They also suffer from similar problems of matching industrial classifications and time-scale coverage as the other series already discussed. These data have been extended to 2003 and are the basic measure of innovatory activity used in the estimation of the equations of E3ME.

The ANBERD database covers only business enterprises, that is, not the public sector, which means that extra work was required before public sector R&D, including defence spending, could be identified. For this type of disaggregation, which was also required in E3ME30, a separate OECD survey based on the Frascati Manual was used, which distinguishes military R&D expenditure for each member state. In addition, the IEA has published annually for the last few years a detailed analysis of OECD states' spending on energy-related R&D, including energy-conservation R&D. This provides detail of large-scale R&D programmes in EU member states funded by national governments.

Unfortunately the data on privately funded energy-saving R&D are partial and incomplete, so it is not possible to present data or results for total energy-saving R&D.

7.2.3.4 GROSS INVESTMENT IN FIXED ASSETS

Detailed data for investment demand (Gross Fixed Capital Formation) in European countries, on both a constant and current price base, are published by Eurostat and are available as part of the OECD's STAN database. E3ME disaggregates investment into the 42 industry sectors used throughout the model, and gaps in the published data were filled using the previous version of the E3ME database (version 3.0). The units are standardized to EUR (current prices) and 2000-valued EUR (constant prices) for all the E3ME countries.

7.2.3.5 METHOD

Investment is a component of GDP, but expenditure on investment and innovation mainly enters E3ME's equations indirectly, by its use in formulating a measure of technical progress. The approach to constructing the measure of technological progress in E3ME is adapted from that of Lee *et al.* (1990). It adopts a direct measure of technological progress (T_t) by using cumulative gross investment, but this is altered using data on R&D expenditure, thus forming a quality-adjusted measure of investment. The equation for T_t is written as

$$T_t = c + ad_t(\tau 1) \tag{7.1}$$

where $d_t(\tau 1)$ satisfies the following recursive formula

$$d_t(\tau 1) = \tau 1 d_{t-1}(\tau 1) + (1 - \tau 1) \log(GI_t + \tau 2 RD_t) \qquad (7.2)$$

where GI_t is the level of gross investment; RD_t is constant price research and development expenditure; $\tau 1$ is a measure of the impact of past quality adjusted investment on the current state of technical advance, while $\tau 2$ is a measure of the weight attached to the level of R&D expenditure.

To initialize the recursive process for d_t, the assumption is made that in the pre-data period the process generating $\log(GI_t)$ is characterized by a random walk. Under this assumption, the first value of d_t can be written as

$$d_0 = \log(GI) \qquad (7.3)$$

where the right-hand side represents the average of gross investment over the first five-year sample period. The values of $\tau 1$ and $\tau 2$ were set at 0.3 and 1.0 respectively, while noting that more sophisticated procedures could have been adopted, such as a grid-search method based on log-likelihood values. The series $d_t(\tau 1)$ is then calculated by working the recursive procedure forward given the initial value, d_0.

In E3ME41, there are two technical progress indicators, one which measures technical progress related to ICT (Information and Communications Technology) investment in the new economy, and one which is related to all other investment. The construction of the two indicators is similar, with investment split up into ICT and non-ICT related investment, and $\tau 2$ set to 0 in the non-ICT investment measure (i.e. all R&D expenditure influences the ICT measure).

The two sets of technical progress indicators appear together in the equations outlined below, and separate long- and short-term parameters are estimated for each one. Due to a lack of data, a single set of indicators is maintained for the EU's newer member states.

7.2.3.6 USE OF THE TECHNOLOGICAL PROGRESS INDICATOR MEASURES

The variables used to represent technological progress enter a variety of equations in E3ME version 4.1, including:

- employment

The technological progress variables are included as part of the implicit production function that lies behind the factor demand equations in E3ME. The effect on employment demand is deemed ambiguous, as this greatly depends on whether the type of technical progress is labour-saving or labour-augmenting. The extra activity of R&D itself, however, is likely to be more labour-intensive than average production in most industries.

- hours worked

The presence of technical progress in determining average hours worked originates in the determination of the optimal number of hours worked as part of the representative firm's cost minimization process. A negative sign is imposed on the coefficients for technological progress, based on the assumption that an increase in investment R&D will improve the efficiency of the capital stock, thus requiring fewer average hours worked for a given number of employees.

- industrial prices

A positive effect was imposed on the technological progress variable to cope with the quality effect that increased investment/R&D is expected to have, that is, the role of product innovation. The effect of process innovation (which would be expected to lower prices) is taken account of by a measure of unit costs, which is a separate variable in the equation.

- export and import volumes

The technical progress indicators are included to help capture the role of innovations in trade performance. This variable could be measured relative to that of competitors, but this has not been implemented in E3ME4.1. (However, if it was only relative technical progress that improved performance and if all countries experienced such progress simultaneously, then there would be no effect on economic growth; this does not appear to be very plausible as a characterization of modern industrial and service economies). The anticipated effects are a positive elasticity for export volumes and a negative elasticity for import volumes.

- export and import prices

The measures of technical progress are included to cope with the quality effect on prices caused by increased levels of investment and R&D, and progress is assumed a priori to have a positive effect on export prices and a negative effect for import prices.

- energy demand

The energy demand equations include both gross investment and R&D spending directly as explanatory variables. These terms are intended to capture the effects of new ways of decreasing energy demand (energy-saving technical progress) and the elimination of older inefficient technologies. This will also take into account the introduction of new energy-saving techniques and methods of energy conservation, and hence is expected to be negative. In particular, technical progress in the industries producing machinery, electronics, and electrical equipment is expected to reduce aggregate energy demand, and technical progress in the motor vehicles industry is expected to reduce the demand for oil, as transport equipment becomes more efficient and alternative energy sources are adopted.

7.2.4 The effects of GHG and energy taxation

7.2.4.1 INTRODUCTION AND ASSUMPTIONS

One of the purposes of E3ME is to provide a consistent and coherent treatment of fiscal policy in relation to greenhouse gas emissions. Figure 7.3 shows how tax rates affect prices and wage rates in the model; the mechanism is the same for taxes on other emissions and energy.

There are inevitably certain simplifying assumptions required in this kind of modelling.

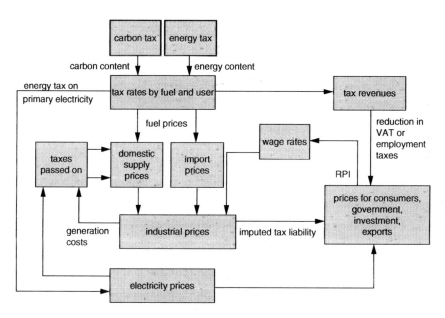

Figure 7.3. The impact of the carbon/energy tax on prices and wage rates

7.2.4.2 FIRST ASSUMPTION

The first assumption is that the effects of the tax in the model are derived entirely through the impact of the tax on fuel prices, and through any use of the subsequent revenues from the tax in reducing other taxes. Other effects are not modelled. For example, if the introduction of such a tax caused the electricity industry to scrap coal-burning plant in advance of what might be expected from the relative price change induced by the tax, this effect would have to be imposed on the model results. The one exception to this rule is the announcement effect of the UK climate change levy (CCL) (see Section 7.6).

All the energy and emission taxes are converted into a consistent set of units (€/toe). These taxes are then added to the costs of the fuels. Tax revenues can be calculated from fuel use; the revenues are reduced according to the fall in use, but rise according to price inflation and any escalator in the tax rates. In the baseline case, effective tax rates are calculated by dividing fuel use by raised revenues; this includes any exemptions and non-payments, which the raw data for tax rates on its own does not.

7.2.4.3 SECOND ASSUMPTION

The second assumption is that imports and domestic production of fuels are taxed according to the energy content of the fuels, but that exports are exempt from the tax coverage. The treatment is assumed to correspond to that presently adopted by the authorities for excise duties imposed on hydrocarbon oils. It is assumed that industries and importers pay the tax, and that it is then passed on in the form of higher fuel prices paid by fuel users. A further assumption is that industrial fuel users may pass on the extra costs implied by the tax in the form of higher prices for goods and services. The increase in final price is a result of the direct and indirect energy content of each commodity distinguished in the model. If the revenues are used to reduce employer tax rates, then industrial employment costs will fall and these reductions in costs are also assumed to be passed on through the industrial system.

7.2.4.4 ETR EFFECTS ON THE E3 SYSTEM

In considering the competitive response of different sectors and companies, there are two important questions to be answered:

- Are the prices of the product set in the world markets or by the producer or in the local markets?

• How flexible is the process of production in responding to an increase in costs?

If the price is fixed in the world market, then no increase in costs arising from an increase in energy taxes can be passed on to final product prices. If the process of production is also fixed (e.g. because the product requires long-lived capital stock), then it might be very expensive to change the technology or move the plant; so all extra costs must be paid out of profits. If the industry or the company is not profitable, then the extra costs could lead to plant closing. However, this is an extreme outcome and most industries and companies have the ability to pass extra costs on to their customers and to change their production process to reduce emissions and avoid some of the increase in taxes.

For these reasons, changes in manufacturing export and import volumes do not give enough information about the effects on competitiveness. The effects on unit costs can be compared to those on export prices to see which sectors have their profit margins squeezed by being forced to accept world prices for their products, while at the same time being unable to avoid increases in their unit costs (see Barker, 1998).

The net effect on industrial and import prices will eventually feed through to consumer prices and will affect relative consumption of goods and services, depending on the carbon/energy content and on their price elasticities. The higher consumer prices will then lead to higher wage claims.

Figure 7.4 shows the effects of these price and wage rate changes on fuel use, CO_2 emissions, and industrial employment. The changes in relative fuel prices as a result of the tax will change fuel use, depending on substitution elasticities. The fuel price increases will be passed on as more general increases in prices, which will cause substitution in consumers' expenditure, in exports, and between imports and domestic production. These changes will feed back to fuel use. CO_2 emissions are derived directly from the use of different fuels. If employment costs are reduced when tax revenues are recycled, then industrial employment will be stimulated directly, with a further indirect effect as labour-intensive goods and services gain in relative price competitiveness.

7.2.5 The effects of the various revenue recycling mechanisms

The COMETR scenarios assume that the ETRs in each country are revenue neutral in each year; in this way, the results presented in this chapter

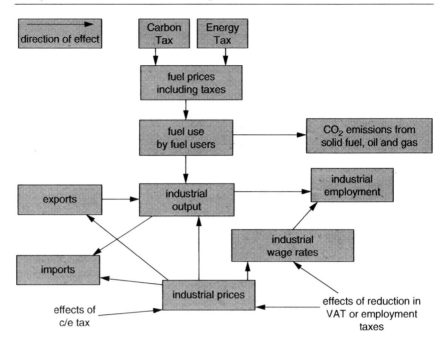

Figure 7.4. The impact of the carbon/energy tax on fuel use, CO_2 emissions and industrial employment

describe a shift in the tax burden to energy use from the more general economy, rather than an increase in the overall tax burden. Of the six countries examined, a variety of methods were used to recycle tax revenues and in some cases a combination of methods was used. These methods are outlined in the list below. For several countries, no explicit revenue recycling mechanisms were put in place, either because the tax reforms were part of such a wide package that it was impossible to determine what the alternative tax arrangements would be, or simply because no provisions were put in place at the time. In these countries, it is assumed that revenue neutrality is achieved through a shift in income taxes. This was judged to be the most non-controversial way of maintaining revenue neutrality, as the ETR revenues are very small when compared to the overall level of income tax revenues.

The revenue recycling methods considered in E3ME were changes in:

- income tax;
- employers' social security contributions;
- employees' social security contributions;

164

- benefit rates;
- investment in energy-saving technology.

The first four methods of revenue recycling relate to the labour market. By changing income tax, the government is directly affecting (nominal) disposable incomes. For example, a reduction in income tax will increase disposable income, all other things being equal. The immediate effects are likely to be a boost to household spending, particularly on luxury goods. Through multiplier effects, there will be further increases in average incomes, as domestic firms require more inputs to meet the extra demand, including labour. In the longer term, a 1 per cent increase in household income will lead to an equivalent 1 per cent increase in overall consumer spending, in line with conventional economic theory. Higher employment and lower unemployment may cause some increases in wages. The case of social security payments is interesting. The effects of changing employees' contributions are almost identical to the effects of changing income tax rates. This makes intuitive sense as the payments by employees are the same, even if the treatment of the tax by government is different (the government sector is largely exogenous in E3ME). The main reason for including this separately from income tax is to simplify the processing in Germany, where there were equal reductions in employers' and employees' contributions.

The initial effect of reducing employers' social security contributions is to lower the cost of labour to firms, and hence should lead to a direct increase in employment. The effects of this will be twofold: first, there will be an increase in average household incomes from higher employment rates; there could also be a slight increase in average wages if unemployment falls. Both effects would lead to an increase in average household incomes and, possibly, a short-term increase in household consumption. The overall inflationary impact will be dependent on the relative strengths of these two effects (e.g. labour-intensive firms will have the largest initial fall in costs and wages will increase faster if the economy is close to full employment), but overall the effects are similar, an increase in average household incomes driving forward consumer spending.

The effects of changing benefit rates are slightly different. If the government increases benefit rates, this increases the disincentive to work. The magnitude of this effect varies across the countries of Europe but, overall, we would expect to see a decrease in employment and labour market participation. In pure economic terms, increasing distortions in the labour market is not seen as a way of increasing productivity and economic output. This must be balanced against equity issues.

By increasing investment in energy-saving products, a government is hoping to achieve a decrease in energy consumption over and above the reduction due to the price effect of higher taxes alone. There will be some other positive effects, however, as investment tends to improve the overall quality and desirability of an industry's output (i.e. increased non-price competitiveness). Previous research has found these effects to be particularly strong in international trade in the long term. More immediately, there will be a boost to industries producing capital goods, such as construction and motor vehicles.

7.2.6 A brief description of E3ME's labour market

E3ME includes a detailed treatment of the labour market with stochastic equations for employment (as a head count), average wages, hours worked, and labour market participation. This plays an important role in the scenarios, particularly in cases where tax revenues are recycled through the labour market. Unemployment is calculated as the difference between employment and the active labour force and is a key explanatory factor in determining wages and household consumption. Unlike many equilibrium models, E3ME does not assume full employment, even in the long run.

7.2.6.1 THE EMPLOYMENT EQUATIONS

Employment is modelled as a total headcount number for each industry and country as a function of industry output, wages, hours worked, technological progress, and energy prices. Industry output is assumed to have a positive effect on employment, while the effect of higher wages and longer working hours is assumed to be negative. The effects of technical progress are ambiguous, as investment may create or replace labour; this will vary between sectors.

7.2.6.2 THE HOURS WORKED EQUATIONS

Hours worked is a simple equation, where average hours worked by industry and country is a function of 'normal hours worked' (expected hours worked based on patterns in other industries and countries) and technological progress. It is assumed the effects of technical progress gradually reduce average hours worked over time as processes become more efficient. The resulting estimate of hours worked is an explanatory variable in the employment equation (see above). Hours worked is defined as an average across all workers in an industry, so incorporates the effects of higher levels of part-time employment in certain countries and industries.

7.2.6.3 THE WAGE EQUATIONS

In E3ME wages are determined by a complex union bargaining system that includes both worker-productivity effects and prices and wage rates in the wider economy. Other important factors include unemployment, tax rates, and cyclical effects. Generally, it is assumed that higher prices and productivity will push up wage rates, but rising unemployment will reduce wages. A single average wage is estimated for each country and sector. The estimates of average wages are a key input to both the employment equations and the price equations in E3ME. In the absence of growing output, rising wages will increase overall unit costs and industry prices. These prices may get passed on to other industries (through the input-output relationships), building up inflationary pressure.

7.2.6.4 THE LABOUR MARKET PARTICIPATION EQUATIONS

Labour market participation is estimated as a rate between 0 and 1 for male and female working-age population. Labour market participation is a function of output, wages, unemployment, and benefit rates. Participation is assumed to be higher when output and wages are growing, but falls when unemployment is high, or benefits create a disincentive to work. In addition, there is a measure of economic structure and the relative size of the service sector of the economy; this has been found to be important in determining female participation rates. The participation rates determine the stock of employment available (by multiplying by working-age population, which is exogenous). This is an important factor in determining unemployment, which in turn feeds into wages and back to labour market participation.

7.3 Processing the COMETR tax data

This section outlines the main stages of processing the tax data set. The aim of this work was to obtain a set of data that could be stored on the E3ME databanks and used in the analysis in the COMETR scenarios.

7.3.1 *Model classification*

Two new classifications were added to the E3ME model so that it could cope with the new tax data. These were the CT (COMETR Tax) classification, and the CR (COMETR Revenue-Recycling) classification.

The CT classification (see Table 7.1) lists all the taxes that are used in the COMETR scenarios. This is not to say that other taxes were ignored, nor that further detail was not used, but it was possible to express all the taxes as an element in this classification. In addition, specific industry information was added at a later stage where it was available. It should be noted that initially there were 26 elements in the CT classification as energy, CO_2 and other emission taxes were separated, but as these taxes are both in terms of EUR/toe energy use (or equivalent measure), they could be simply added together and there was no information gained from keeping them separate. In addition, much of the tax revenue data combined energy, CO_2 and other emission tax receipts, so this proved to be a more efficient use of data. The original classification was expanded to include petrol and diesel separately rather than having a single motor spirit entry.

The CR classification (see Table 7.2) handles the various revenue recycling mechanisms employed in the ETRs considered in COMETR.

Table 7.1. The CT classification

1. Industry energy and CO_2 tax: Coal
2. Industry energy and CO_2 tax: Oil (heating)
3. Industry energy and CO_2 tax: Gas
4. Industry energy and CO_2 tax: Electricity
5. Industry energy and CO_2 tax: Petrol
6. Industry energy and CO_2 tax: Diesel
7. Household energy and CO_2 tax: Coal
8. Household energy and CO_2 tax: Oil (heating)
9. Household energy and CO_2 tax: Gas
10. Household energy and CO_2 tax: Electricity
11. Household energy and CO_2 tax: Petrol
12. Household energy and CO_2 tax: Diesel

Source: Cambridge Econometrics.

Table 7.2. The CR classification

1. Reduction in income taxes
2. Reduction in employers' social security contributions
3. Reduction in employees' social security contributions
4. Increase in state benefits (inc pensions)
5. Additional government investment

Source: Cambridge Econometrics.

7.3.2 Rates, revenues, and recycling

7.3.2.1 THREE MAIN TYPES OF INPUT DATA WERE REQUIRED

E3ME's energy and environment databanks already contain detailed data on fuel demand, by 19 fuel users and 12 fuels (source: IEA). This forms the base for the tax. The tax rates were compiled, based around the classifications outlined above.

E3ME used the tax rates to estimate the increase in the cost of fuels as a result of the ETR. This was fed into the energy submodel, the fuel demand equations, and then into the rest of the model.

Tax revenues were also required. It would be intuitive to say that tax revenues should be equal to tax rates * fuel use, but this misses out an important part of the modelling (and indeed the focus of one scenario), in that the full tax rates are rarely paid at a macro level. This may be due to special exemptions to specific industries or households, or the failure of central governments to collect tax. In the COMETR reference case, the effective tax rates were determined by the tax revenues divided by fuel use. While this makes some attempt to take into account exemptions and non-payments, it turned out that the data for tax revenues were generally much harder to obtain and are published at a more aggregated level, so assumptions had to be made on a case-by-case basis. See the following sections for more details.

The revenue recycling is an important part of the modelling in E3ME, and in some cases has a larger economic impact than the energy taxes. Therefore it is important that the scenario results demonstrate the most accurate profile of revenue recycling possible, and are not biased by any changes in the overall level of taxation. To achieve this target, the revenue recycling data were converted to shares of revenue received and the shares were set to sum to 1, ensuring that tax receipts equal revenue recycling payments.

All three data sets were obtained in the form of annual time series covering the period 1994–2004. Gaps in the time series (e.g. in 2004) were estimated using linear interpolation or extrapolation based on projections of fuel use and Cambridge Econometrics' custom software algorithms.

7.3.2.2 STANDARDIZED UNITS WERE USED TO STORE THE DATA

The main unit of energy data in E3ME is thousands of tonnes of oil equivalent (th toe) and the economic variables are stored and calculated in millions of EUR (tax data are held at current prices, then deflators used to obtain 2000-based series). The units chosen for the COMETR classifications are consistent with this, namely being:

- tax rates are held in EUR / th toe;
- tax revenues are held in millions of EUR;
- revenue recycling is stored as shares and used to make changes to other taxes in millions of EUR or investment (millions of 2000-based EUR).

In cases where the data provided were in national currency, the exchange rates on E3ME's economic time-series databank were used to convert the data to EUR.

7.3.2.3 THE MAIN CONVERSION FACTORS USED IN THE DATA PROCESSING

These were the main conversion factors used for the energy classifications:

- 1 toe = 41.868GJ
- 1 toe = 11.63 MWh
- 1 tonne of oil = 1,192 litres
- 1 tonne of petrol = 1,362 litres
- 1 tonne of diesel = 1,203 litres
- coal produces 25.4GJ/tonne
- oil produces 43.5GJ/tonne
- gas produces 35.6MJ/CUM
- petrol produces 44.8GJ/tonne
- diesel produces 43.3GJ/tonne

7.3.3 *Software inputs*

The data were received in a single Microsoft Excel spreadsheet.

The data processing was done using the Ox software package (see <http://www.doornik.com/products.html#Ox>). Ox is a flexible matrix-based software package that has commands similar to those in standard C++ in construct. Ox was used to read in raw data from the spreadsheets, process it, and save it to the E3ME databanks. There was also some interaction between Ox and Visual Basic in accessing the spreadsheet data efficiently and Cambridge Econometrics has its own library of custom Ox software to aid with the processing.

E3ME is programmed in Fortran and controlled by the IDIOM software package. E3ME's direct-access databanks are Fortran-based, but can also be accessed by Ox.

7.3.4 *Processing the individual countries*

7.3.4.1 INTRODUCTION

This section outlines the main steps and assumptions made in order to process the data for the individual countries as accurately as possible. All data were converted into the units given above using the converters described above. The data for tax rates were usually available, so it was not necessary to make many assumptions during the processing. However, this was not generally the case with revenues and revenue recycling methods. Much of this work drew from the analysis carried out by other partners in the COMETR project.

7.3.4.2 DENMARK

Denmark is different from the other countries in that it makes a clear distinction between 'light' and 'heavy' industry and charges different tax rates to each one. This poses a problem for models such as E3ME that expect a single value. The final methodology counted all industry as heavy, as the majority of emissions fall in this category, but without deductions.

The tax revenues for Denmark were split between energy and CO_2 taxes. Although the two were eventually aggregated, this was useful in allocating between fuels and sectors. The energy tax was split by fuel and was not applied to industry (except motor fuels), so this part was straightforward (fuels were shared out assuming that exemptions are equal for households and industry). CO_2 taxes were slightly more problematic in that only a national total was available and so these had to be split between households and industry, and also between the different fuel types. With no other information available the sharing was done using fuel tax * fuel uses, that is, the exemptions are the same across households and users. It is unfortunate that no additional information was available to allow this assumption to be relaxed.

Taxes that existed before ETR commenced in 1992 were not included in the analysis. Existing taxes were defined as ones that existed in 1991 and were subtracted from total tax rates to calculate just the ETR part. Likewise, a similar share of the revenues was removed and attributed to existing taxes.

7.3.4.3 GERMANY

Although the tax revenue data for Germany were relatively detailed compared to other countries, the data did not make the distinction between industry and households. Therefore it was necessary to make the following assumptions:

- for motor spirit and electricity, households have the same exemptions as industry;
- for any other fuels, households always pay the full rate.

As household consumption of fuels other than middle distillates and electricity is very low, the second of these assumptions is not particularly important. However, particularly in the case of electricity, assuming the same exemptions are available for households and industry may not be realistic.

Revenue recycling in Germany was relatively complex compared to the other countries, in that three main methods were used. Working on the basis that the cuts in social security contributions were shared equally between employers and employees, the shares were calculated by dividing the extra investment by total revenues, and equally splitting the rest between employers' and employees' social security contributions, so that the shares summed to 1. Existing tax rates were subtracted from the totals, allowing an analysis of just the ETR component of the tax.

7.3.4.4 FINLAND

These data included seven fuels that were aggregated to the fuels in the CT classification. Interestingly, no revenues for gas were included; it is possible that these data were part of light fuel oil but with no information to go on, it was decided to add this separately by calculating tax rate * fuel use (i.e. assuming zero exemptions). As gas use is quite minor in Finland, this assumption should have had little overall bearing on results.

The other major limitation with the Finnish data was that there was no distinction between revenues from industry and households. After reviewing the available literature, it was assumed that households had no exemptions and always paid the full charge. Exemptions for industry were then calculated as total revenue—(industry tax * industry fuel use). Existing taxes in Finland (those that existed in 1996) were subtracted from the totals and were not considered to be part of the 1990s ETR.

There was no specific policy covering revenue recycling in Finland, so it was assumed that all revenues from ETR were compensated for by reductions in income tax. The only special industry exemption to take into account in Finland was that for very large firms, 10–12 in number, mainly in the paper and pulp industry. However, the scale of the exemption was up to 85 per cent and this is a very large sector in Finland, so an effort was made to include this. According to the Finnish Forestries Industries Federation, the five largest firms made up 85 per cent of turnover in the sector. Therefore it was assumed that the ten largest firms made up 90 per cent of the sector and energy use. Given these assumptions, it seemed reasonable to ignore the relatively small threshold of €50,000 below which all tax is paid, and simply reduce payments from this sector by 100 * 0.85 * 0.9 per cent. Consequently, the pulp and paper sector only paid 23.5 per cent of the tax.

7.3.4.5 THE NETHERLANDS

Much of the processing for the tax rates in The Netherlands was fairly simple in nature, with an average tax rate calculated for heating oil from gas oil and kerosene. Excise duties were not counted as part of the ETR, and any taxes that existed before 1998 were not counted as part of the modelled ETR.

There were major difficulties in estimating industry tax rates and revenues for gas and electricity, however. Energy tax rates for these fuels in The Netherlands are dependent on the size of user (in terms of fuel consumption); in the case of households, it was assumed that all users fell into the smallest category, but this was not a valid assumption for industry. After extensive searching, no relevant data were found for firm size in terms of energy use; while it would have been possible to use the closest data (firm size based on employment or turnover), there was no guarantee that this would have been any more accurate than using a single estimate (and this may also have introduced bias between sectors), so a single category was chosen. This was 50,000–10m kWh of electricity and 170,000–1m cubic metres of gas. In the scenarios with no exemptions, the highest tax rates were used.

A further complication in processing The Netherlands data was that the revenues were disaggregated into just two categories: energy tax and other environmental taxes. As there was no specific mention of exemptions on fuels with the simpler taxes (coal and oil), these revenues were assumed to be correct, and the more complex systems (gas and electricity) were scaled so that the revenues in E3ME matched the published total.

As there was no specific treatment of revenue recycling in the Dutch ETR, it was assumed that the alternative was higher direct income taxes.

7.3.4.6 SWEDEN

The approach for Sweden was very different from the other countries. Statistics Sweden publishes a very detailed set of revenues from environmental taxes, disaggregated by NACE two-digit sector. It was decided to make use of these data, rather than rely on E3ME to estimate the sectoral revenues. This means that a separate rate of exemption was available for each industry. This does assume, however, that the industry exemptions are independent of the fuel mix to that industry—that is, the same exemptions will apply to coal, gas, and renewable energy (although this of course does not mean that the actual tax rates do not vary by fuel type).

As a result of this, the tax revenue data for Sweden were stored as a 19 x 11 matrix (19 fuel users and 11 years) rather than a 12 x 11 matrix (the 12 CT categories and 11 years). E3ME required specific adjustments to cope with this.

Finally, the time series for Swedish tax revenues were extrapolated to include 2003–4. This was done by assuming a linear relationship between the tax revenues and the given tax rates multiplied by projected fuel use.

With no further information, it was assumed that all revenues from ETR were recycled in the form of reduced income taxes.

7.3.4.7 THE UNITED KINGDOM

The climate change levy (CCL) rates were easily obtainable in the UK, but the revenues from the tax are only available as an aggregate for the UK. Individual industries that do not pay the CCL were exempted from the tax during the modelling stage (based on CE's fuel user classification), but the assumption was that exemptions were equal across fuels, so the revenues were allocated to fuels in line with total consumption of that fuel.

By using the data for revenues, the negotiated agreements were taken into account. The scenarios principally looked at the price effects of the CCL and not energy savings made in response to the negotiated agreements; however, the announcement effect on energy demand by other final users was taken into account. Analysis suggests these negotiated agreement effects are not insignificant, so results published here may be understating the overall drop in fuel demand and emissions resulting from the CCL.

As the CCL is a completely new tax in the UK, there was no issue about what counted as part of the 1990s ETR and what was already in place. Revenue recycling was assumed to have occurred completely through the effects of reducing employers' social security contributions.

7.3.4.8 SLOVENIA

Although the CO_2 tax in Slovenia was not, strictly speaking, part of an ETR, it was included in the baseline scenario to give an example of environmental taxation in the new member states. For the purpose of the modelling, it was assumed that the revenues were recycled through reductions in direct income taxes.

It was very difficult to define the CO_2 tax in Slovenia, with environmental taxes often being bundled with other taxes and different data sources giving conflicting stories. Following consultation, it was decided that only a tax on natural gas consumption should be included. With no data for revenues, it was necessary to assume zero exemptions in all the scenarios, and to estimate revenues as tax rates multiplied by fuel use.

The CO_2 tax was not applied to the power generation sector, so it was excluded in the scenarios.

7.3.5 Revenue recycling methods

Table 7.3 illustrates the mechanisms used for recycling revenue. Section 7.3.6 describes these methods in more detail and outlines the expected results in each case.

7.3.6 Key assumptions made in the study

7.3.6.1 MODELLING ASSUMPTIONS

Unless otherwise stated, all the modelling follows the same assumptions as the E3ME model. These are documented in the model manual, which is available online at <http://www.camecon-e3memanual.com/>.

In addition, the following assumptions were made:

• All of the taxes are revenue-neutral.

Although there are cases where the ETRs are not designed or intended to be revenue neutral, this was imposed in the modelling so that the results indicate the effects of a shift in the tax burden rather than an overall increase or decrease in the tax burden. In cases where the data did not

Table 7.3. Revenue recycling by country (million €)

		1994	1995	1996	1997	1998	1999	2000	2001	2002	2003	2004
Denmark	Income tax	0	0	0	0	0	0	0	0	0	0	0
	Social security contributions	271	615	943	1,032	1,311	1,702	1,820	1,898	2,044	2,134	2,140
	Investment	0	0	8	13	17	28	26	13	0	0	0
Germany	Income tax	0	0	0	0	0	1,952	4,002	5,698	6,951	9,009	9,181
	Social security contributions	0	0	0	0	0	1,952	4,002	5,698	6,951	9,009	9,181
	Investment	0	0	0	0	0	201	197	304	183	182	204
Finland	Income tax	0	0	0	373	614	700	685	721	722	895	894
	Social security contributions	0	0	0	0	0	0	0	0	0	0	0
	Investment	0	0	0	0	0	0	0	0	0	0	0
Sweden	Income tax	−124	−50	349	657	1,550	1,570	1,852	1,741	1,992	2,395	2,585
	Social security contributions	0	0	0	0	0	0	0	0	0	0	0
	Investment	0	0	0	0	0	0	0	0	0	0	0
United Kingdom	Income Tax	0	0	0	0	0	0	0	0	0	0	0
	Social security contributions	0	0	0	0	0	0	0	540	1,372	1,134	1,201
	Investment	0	0	0	0	0	0	0	0	0	0	0
Slovenia	Income tax	0	0	0	0	0	16	43	57	46	45	45
	Social security contributions	0	0	0	0	0	0	0	0	0	0	0
	Investment	0	0	0	0	0	0	0	0	0	0	0

Source: Cambridge Econometrics.

Table 7.4. ETR as a percentage of GDP, 2004

	DK	DE	NL	FI	SW	UK	SI
ETR €m	2,140	18,547	2,287	894	2,585	1,200	45
GDP €m	197,222	2,207,200	489,854	151,935	281,124	1,733,603	26,232
ETR as a % GDP	1.08	0.84	0.47	0.59	0.92	0.07	0.17

Source: Cambridge Econometrics.

support this, shares were used to scale the revenue recycling to match the tax revenues. In cases where there was no clear method of revenue recycling (Finland and Sweden), it was assumed that environmental taxes were an alternative to higher direct income taxes.

• Tax rates in the non-ETR countries and economic activity outside the EU were assumed to remain constant in the scenarios.

7.3.6.2 DATA ASSUMPTIONS

This section summarizes the main assumptions made during the data processing in order to get a complete data set, and reflects our attempts to make best use of the information available.

Where detailed tax revenues were missing, typically the aggregates were shared out using shares of (fuel tax * fuel use), assuming exemptions were similar across fuels or industry. If the literature suggested that there were no exemptions in a particular group, then this total was entered into the data and the remainder of the aggregate tax receipts shared out. Where time series did not cover all of the period 1994–2004, linear interpolation or extrapolation based on fuel use was used to estimate missing values. Tax rates were assumed to remain constant when no information was available. Tax rates were assumed to remain constant in real terms over the forecast period. Table 7.4 shows that ETR as a percentage of GDP in 2004 was less than 1.1 per cent for all the ETR countries. There are noticeable differences between the ETR countries; in the UK, the ETR accounts for just 0.07 per cent of GDP, compared to 1.08 per cent in Denmark, 0.92 per cent in Sweden, and 0.84 per cent in Germany.

7.4 Scenarios specified to model ETR

The ETR effects were modelled using scenarios that consider the various components of the green tax reform packages described above.

Table 7.5. COMETR scenarios

Scenario	ETR	Revenue recycling	Exemptions
1. Reference (R)	No	No	N/A
2. Baseline (B)	Yes	Yes	Yes
3. Exemptions (E)	Yes	No	Yes
4. Compensation (C)	No	Yes	N/A

Source: Cambridge Econometrics.

7.4.1 Description of ETR scenarios

The following scenarios, shown in Table 7.5, were generated by E3ME over the period 1994 to 2012 so that the projection period includes Phase 2 of the EU ETS:

  (i) The reference case (R) A counterfactual projection without the ETR, but including current and expected developments in the EU economy, for example the EU ETS.
 (ii) The baseline case (B) An endogenous solution of E3ME over the period 1994–2012. This scenario included the ETR in each member state covered by the project, exemptions or special treatment for the industries most affected, and compensating reductions in other taxes. This scenario was calibrated closely to the observed outcome through using historical data which include the effects of ETR implementation.
(iii) The tax with no exemptions and special treatment case (E) (but with compensating measures).
 (iv) The compensating reduction in another tax on its own (C).

These scenarios allow the ETR to be decomposed country by country into three components: the full tax, the exemptions, and compensation via recycling.

7.4.2 Creating the baseline solution

The baseline is an endogenous model solution of E3ME that fully covers the period 1994–2012 annually. The baseline solution is calibrated to be consistent with a combination of historical data and forecast. The historical part of the solution, which is used for the ex-post analysis of 1994–2002 (or 2003 where the data were available at the start of the project), comes from E3ME's historical databanks. The main sources for these are the OECD, Eurostat, and the IEA. These data include the effects of ETR in the 1990s, and any accompanying exemptions and revenue recycling methods. The forecast part of the baseline solution, which is used for the ex-ante analysis, is derived from a combination of DG TREN's Energy and Transport Trends to 2030 (with emissions data published by the EEA), and the IEA's energy price assumptions.

While the DG TREN forecast provided a consistent set of forecasts for energy demand and economic activity, the assumptions underlying the forecast (namely energy prices, world growth, and the ETS) were outdated. To compensate for this, E3ME was calibrated to meet the DG TREN forecast, and then solved again with a different set of assumptions for energy prices which take into account the shock to oil prices over 2004–5, and the ETS allowance price. The energy price assumptions are important because they set the ratio of environmental tax to total fuel cost (i.e. the difference between the scenarios) which determines the scale of the effects. However, it was not possible to obtain more recent forecasts for world GDP growth that were consistent with the other inputs.

The COMETR project was the first time that E3ME was solved endogenously over its historical period.

7.5 Estimation of competitiveness effects

7.5.1 *Estimated competitiveness effects in COMETR sectors*

Five sectors were chosen at the NACE two-digit level, in line with the other COMETR analyses. One of these sectors covers two E3ME sectors (the E3ME pharmaceuticals and chemicals nes sectors make up NACE code 24), so both of these sectors were included. The full list is shown in Table 7.6.

The six countries that followed a path of environmental tax reform in the 1990s are:

- Denmark;
- Germany;
- The Netherlands;
- Finland;
- Sweden;
- The UK.

179

Table 7.6. Sector classifications

NACE 2 digit code	Full Description	E3ME sectors	E3ME Description
15	Manufacture of food and food products	5	Food, drink and tobacco
21	Manufacture of pulp, paper and paper products	7	Wood and paper
24	Manufacture of chemicals and chemical products	10	Pharmaceuticals
		11	Chemicals nes
26	Manufacture of other non-metallic mineral products	13	Non-metallic mineral products
27	Manufacture of basic metals	14	Basic metals

Source: Cambridge Econometrics.

For Slovenia, the CO_2 tax, although not strictly part of an ETR, was included in the baseline scenario to give an example of environmental taxation in the new member states.

7.5.2 Modelling results

E3ME was extended to include the ten countries that became members of the European Union in 2005 (Bulgaria and Romania were added at a later date). New data were gathered and a complete new set of equations was estimated for the entire EU25.

To cope with the shorter time series of available data for the new member states, a shrinkage estimation technique was developed and used to estimate the long-term parameters in these countries.

The parameter estimates determine the behavioural relationships within E3ME and can give an early indication of the scenario results.

Non-price competitiveness is modelled through E3ME's technical progress indicators, which provide a measure of product quality. Both price and non-price competitiveness effects were found to be very important in the trade equations, particularly when considering trade within the European single market.

7.6 The effects of selected ETRs, using E3ME, 1995–2012

7.6.1 Model results: overall effects of ETR

This section compares the results for the baseline case against the reference case. In summary, this illustrates the difference between what did happen and what would have happened had there been no ETR (with both cases projected to 2012). The exception to this was that revenue neutrality was assumed in each case through the revenue recycling mechanisms. Exemptions, non-payments, and negotiated agreements were included as accurately as possible as they happened, subject to the total revenues matching the published figures in each case. Therefore the data used for this analysis were the tax revenues collected as described above. The taxes were not assumed to have any effect other than to increase energy prices (e.g. there are no extra awareness effects), with the exception of the CCL announcement effect in the UK.

This account of effects is focused on the outcome for the environment and economic activity. Chapter 8 reports the results for international competitiveness in the context of carbon leakage.

7.6.1.1 ENERGY DEMAND

As the taxes included in the analysis increased fuel prices, we would expect the primary effect to be a reduction in the demand for energy. The scale of the reduction will depend on tax rates, on how they are applied to the various fuels and fuel user groups, on how easy it is for fuel users to substitute between different fuel types and non-fuel inputs, and on the scale of the secondary effects from resulting changes in economic activity.

All the six countries show a reduction in fuel demand from the ETR (see Figure 7.5). In most cases, the reduction in fuel demand was in the

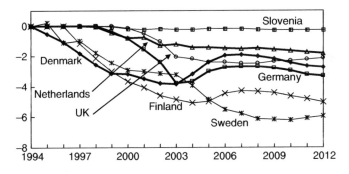

Figure 7.5. The effect of ETR on total fuel demand

Note: % difference is the difference between the base case and the counterfactual reference case.

Source: Cambridge Econometrics.

region of 4 per cent, although it was slightly larger in Sweden and Finland than in the other countries. A key feature of the results is the recovery in fuel demand in several of the examined countries over 2004–5 in the baseline case relative to the reference case, due to higher world energy prices, included in both the baseline and reference cases. In most of the ETRs, environmental taxes were not raised in line with fuel prices (and in some cases may have been reduced), implying a reduction in the relative change in fuel prices. For example, a tax that doubled the price of oil for households in 2003 may only have increased it by 50 per cent in 2005. Consequently the change in fuel demand becomes less in these years. After 2004, the environmental taxes are assumed to increase in line with the consumer price index (not energy prices). Results after 2004 are therefore mainly a reflection of changing energy prices, and the dynamic and lagged effects of the ETRs.

7.6.1.2 GHG EMISSIONS

We would expect to see a reduction in atmospheric emissions from lower fuel consumption, but total emissions also depend on the relative consumption levels of each fuel type. For example, a tax system that encourages the use of coal is likely to produce higher emissions than one which encourages the use of natural gas or bio-fuels. E3ME includes explicit equations for fuel shares of hard coal, fuel oil, natural gas, and electricity. Assumptions are made about the other fuel types, linking them to the closest modelled alternative (e.g. other coal is linked to hard coal, crude oil to fuel oil). For middle distillates (petrol, diesel, etc.), demand is linked to total fuel demand by that sector. The reason for this is that demand for these fuels is dominated by the transport sectors. These sectors do not generally use any other fuels, so fuel share equations are not required.

The scenario results show that there are reductions in greenhouse gases (GHGs) in all six ETR countries from the ETRs (see Figure 7.6). This is consistent with national policies to meet the European Burden Sharing Agreement targets, by which EU countries have agreed to emit a specified level of GHGs over the period 2008–12. The effects closely follow the results for total fuel consumption, with the largest reductions occurring in countries with the highest tax rates. The largest reductions in emissions occur in Finland and Sweden. It should be noted that, in most cases, the fall in emissions is relatively larger than the fall in fuel demand, indicating that tax policies are efficient at reducing emissions.

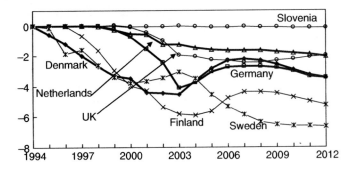

Figure 7.6. The effect of ETR on GHG emissions

Note: % difference is the difference between the base case and the counterfactual reference case.

Source: Cambridge Econometrics.

7.6.1.3 GDP

As a general rule, the effects of the ETR are positive on economic activity, depending on how revenues from the environmental taxes are recycled. This is because tax distortions in the labour market are reduced in the revenue recycling measures. Lowering labour taxes encourages employers to hire more staff (in the case of employers' contributions) and encourages more people to join the labour force (in the case of income taxes). Unlike some economic models, E3ME does not assume that economies operate optimally and shifts in the tax burden will therefore not necessarily have a negative effect on output. However, it is likely that there will be short-term transition costs, so the gains may not be immediate. All six of the ETR countries have an increase in GDP as a result of the ETR (see Figure 7.7). In Sweden, the effects take slightly longer to come through, as the very large increase in household electricity taxes depresses real incomes in the short run. Finland has a short-term boost to GDP from the effects of the taxes on fuel demand, because a reduction in the demand for imported fuel improves the country's trade balance.

7.6.1.4 EMPLOYMENT EFFECTS

The ETR caused employment in some of the ETR countries to increase by as much as 0.5 per cent. Employment increases because the revenue from the ETR is used to reduce employers' social contributions, meaning labour costs are reduced, and firms are able to increase their labour force. This is the case for Denmark, Germany, and the UK. In Denmark, the ETR has an immediate effect on the level of employment, which

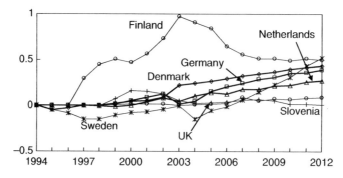

Figure 7.7. The effect of ETR on GDP

Note: % difference is the difference between the base case and the counterfactual reference case.

Source: Cambridge Econometrics.

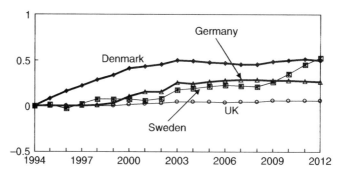

Figure 7.8. The effect of ETR on employment

Note: % difference is the difference between the base case and the counterfactual reference case.

Source: Cambridge Econometrics.

remains nearly 0.5 per cent higher throughout the modelling period (see Figure 7.8). In Germany, a more modest increase in employment is recorded of approximately 0.2 per cent increased employment against the reference case. However, in the UK, the change in employment is small, as the revenue recycled to reduce social security contributions was much smaller. Interestingly, employment in Sweden was higher due to the ETR despite revenues being used to reduce income tax and not social security contributions. This is because the increase in GDP as a result of the ETR caused employment to increase slightly compared to the reference case.

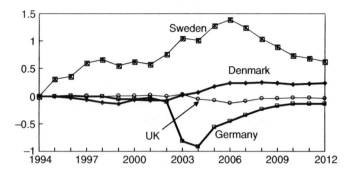

Figure 7.9. The effect of ETR on investment

Note: % difference is the difference between the base case and the counterfactual reference case.

Source: Cambridge Econometrics.

7.6.1.5 EFFECT ON INVESTMENT

Investment levels remain unchanged for most of the ETR countries. Figure 7.9 shows that investment levels are virtually unchanged between the baseline and the reference case for the UK and Denmark (similarly for Finland, Slovenia, and The Netherlands, although this is not shown). In Sweden, investment increases as a result of the ETR by nearly 1.5 per cent in 2006. Conversely, in Germany, investment falls as a result of the ETR. This is due primarily to an increase in energy prices as a factor cost, but also due to the relative cost of capital to labour, with lower social security rates favouring labour. However, by 2012, the levels of investment are broadly similar between the baseline and the reference scenarios.

7.6.1.6 INFLATIONARY EFFECTS

As the ETRs result in higher fuel prices, it is intuitive that there should be an increase in the overall price level. The degree of this is likely to be dependent on the scale of the increase in fuel costs, how easy it is for industry and consumers to switch between fuels to cheaper alternatives (and non-energy inputs), and how much of the cost is passed on by industry to consumers (this is dependent on the level of competition in the industry, which is estimated econometrically for each country and sector). It should also be noted that revenue recycling may have a depressing effect on inflation, as in Germany, when the revenues are recycled through reductions in employers' social security contributions (i.e. labour costs).

185

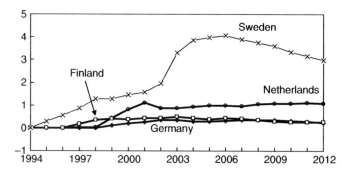

Figure 7.10. Effects on the consumer price index

Note: % difference is the difference between the base case and the counterfactual reference case.

Source: Cambridge Econometrics

In Denmark and in the UK, there were no significant increases in the overall price index. In the UK, this is because the tax is relatively small and was compensated with slightly cheaper labour costs. In Denmark, the tax was larger, but was again compensated with lower labour costs.

The measure of inflation shown in Figure 7.10, the consumer price index (including housing costs), will record a larger increase in cases where taxes are levied on households rather than on industry. The reason for this is that the consumer price index is a weighted average of the price of consumer products, including energy. In the cases where the tax is levied on households, the whole tax is reflected in the consumer price index, rather than just the share that is passed on by industry. Therefore, it is not unexpected that the largest increases are in The Netherlands and, in particular, in Sweden. The largest increase in wages by far is in Sweden, where there is an average increase of 3–4 per cent (roughly the same as the increase in the CPI). In the other countries, the effect on wages is in the region 0–0.5 per cent, with the largest increases in The Netherlands and Finland.

7.6.2 Effects of exemptions

It is not always easy to specify, in terms of inputs to the modelling, the legal or procedural exemptions included in environmental tax laws or their effects on tax revenues. For example, there may be little difference between a lower tax rate and an exemption from a higher tax rate. It is almost impossible to obtain data for lost revenues from exemptions, and the data that do exist are likely to include other non-payments. In the scenario with no exemptions, we have defined a tax with no exemptions as the full tax rate, and revenues from this tax are calculated as tax rate times fuel use. This is the opposite case to the baseline scenario, where tax rates are defined as revenues divided by fuel use. This approach makes the implicit assumption that all non-payments are exemptions and not, for example, due to tax evasion.

Generally, the effects are fairly linear, in that the higher tax rates cause a larger decrease in fuel use and emissions, and this feeds through to the wider economy. Usually, the effects of the exemptions are quite small, however. The largest effect is in The Netherlands, because the tiered electricity and gas rates to business are assumed to be exemptions, and therefore, when these are removed, industry pays the highest tax rates and the effects on fuel demand are much greater.

7.6.3 Isolating the effects of taxes

This section looks briefly at one of the alternative scenarios that is designed to identify the effects of energy taxes on their own without any revenue recycling (see Figure 7.11). We would expect higher energy prices on their own to have a negative impact on GDP, although in one country, Finland, GDP increases due to energy imports (particularly oil and gas) falling.

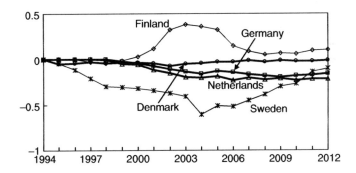

Figure 7.11. Effects of ETR on GDP without revenue recycling

Note: % difference is the difference between the base case and the counterfactual reference case.

Source: Cambridge Econometrics.

The assumptions underlying this analysis are stylized and therefore not necessarily realistic. As national governments are not using the extra revenues to reduce labour taxes or increase spending, it is assumed that revenues are used simply to reduce national debt. As the government sector and interest rates are exogenous in E3ME, this is effectively saying that there is an increase in the overall tax burden and the revenues raised by government are leaving the system. The overall effect (higher fuel prices with no compensating measures) is equivalent to an increase in world energy prices in countries that have no domestic oil and gas resources.

7.6.4 Sensitivity of the results to key inputs

It is important to check the robustness of the modelling results to changes in key inputs. As the scenarios focus on changing patterns of energy use, one of the most important inputs is energy prices. Previous results have already illustrated how the influence of environmental taxes became less as world energy prices increased over the period 2004–5, and it cannot be assumed that the effects will not diminish at other times and under other circumstances.

To test the sensitivity of the results to changes in energy prices, two additional sets of model runs (Baseline and Reference cases) were created, and the results were examined for major differences from the main set of results (see Tables 7.7 and 7.8).

These extra model runs were identical to the main project scenarios except that the energy prices fed into E3ME's energy submodel were reduced by 10 per cent in one set (low) and increased by 10 per cent in the other (high). This increase or decrease was applied to all energy

Table 7.7. High fuel prices: total fuel demand (% difference of baseline from reference case)

	1994	1998	2002	2006	2010
DK	0.00	−2.56	−3.76	−1.93	−2.44
DE	0.00	0.00	−2.36	−2.76	−3.02
NL	0.00	−0.07	−1.24	−1.46	−1.64
FI	0.00	−2.09	−4.53	−4.37	−4.54
SW	0.00	−1.78	−3.12	−5.40	−6.09
UK	0.00	0.00	−1.05	−2.40	−2.29
SI	0.00	0.00	−0.17	−0.19	−0.19

Source: Cambridge Econometrics.

Table 7.8. Low fuel prices: total fuel demand (% difference of baseline from reference case)

	1994	1998	2002	2006	2010
DK	0.00	−2.56	−3.78	−1.92	−2.43
DE	0.00	0.00	−2.36	−2.75	−2.91
NL	0.00	−0.07	−1.27	−1.46	−1.64
FI	0.00	−2.14	−4.65	−4.50	−4.70
SW	0.00	−1.78	−3.16	−5.60	−6.36
UK	0.00	0.00	−1.02	−2.35	−2.25
SI	0.00	0.00	−0.17	−0.19	−0.26

Source: Cambridge Econometrics.

products in all countries in each year over the period 1994–2012, with a 1 pp (percentage point) increase or decrease in each year from 1994 to 2004. The counterfactual runs were for the historical data available for 1994 to 2003. No explicit reason was given for the increase in energy prices and these extra model runs cannot be considered as well-defined scenarios, because other exogenous model inputs (e.g. world growth, ETS allowance prices) were unchanged.

The results of this exercise show that, as expected, the impact of energy taxes is higher in the low-price scenario in the two countries with the largest effects, Finland and Sweden. This is because taxes have a larger relative effect when energy prices excluding taxes are lower. The difference in the reduction in fuel demand ranges from zero to around 0.25 pp in 2012 between the high- and low-price scenarios. The true result, reported previously, sits halfway in between. This means that up to 2 per cent of the reported difference could be attributable to a 10 per cent change in the oil price in either direction.

There is virtually no difference in the largest countries, Germany and the UK. If anything, taxes have a larger effect when there is a higher oil price. It is not obviously clear why this is, but it may well be a result of the relatively small scale of the taxes, and the fact that they were only introduced later in the period, when oil prices were already higher in these scenarios. In The Netherlands, the bulk of the tax fell on domestic electricity use, in which prices changed much less as a result of the change in oil price (i.e. not much of the change was passed on to consumers and less than in Sweden), so we would not expect to see much change in the results. It should also be noted that the UK and The Netherlands are oil-producing countries and so will react slightly differently to a change in world oil prices compared with other countries.

7.6.5 The effects of ETR on competitiveness in individual sectors

A direct method for investigating the impact on competitiveness is to determine whether or not the introduction of ETR caused export and import levels to change in the ETR countries. Figure 7.12 shows the effects of the ETR on aggregate exports by country.

In the largest two ETR countries, Germany and the UK, the ETR has very little effect on exports; similarly, there was little effect on exports in Finland and Sweden. However, Denmark and The Netherlands see a small increase in exports over the period. In the case of Denmark, this is the result of lower labour costs, as revenues are recycled by reducing the social security contribution.

Sweden witnesses, as Figure 7.13 shows, the largest increase in imports. Aggregate imports are 0.8 per cent higher as a result of the ETR. It is unlikely that this represents a reduction in competitiveness in the domestic market for home producers; rather, given that GDP increases by 0.5 per cent in Sweden and that revenue is recycled into reducing income taxes, it is more probable that imports increase as a result of increased consumer spending. Both Germany and the UK, the two largest ETR economies, see little change in imports as a result of the ETR.

The COMETR project focuses on four of the most energy-intensive E3ME sectors, plus food and pharmaceuticals to provide a comparison. These are defined in Table 7.9 below at the NACE two-digit level.

Before analysing the individual sectors, it is worth looking at what proportion of inputs comes from the energy sectors. Table 7.10 shows the

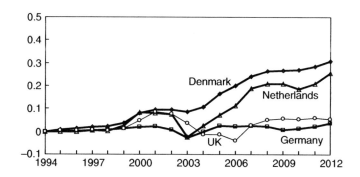

Figure 7.12. The effect of ETR on exports

Note: % difference is the difference between the base case and the counterfactual reference case.

Source: Cambridge Econometrics.

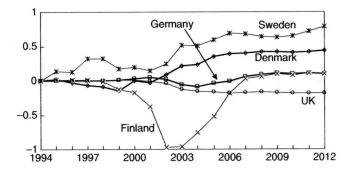

Figure 7.13. The effect of ETR on imports

Note: % difference is the difference between the base case and the counterfactual reference case.

Source: Cambridge Econometrics.

Table 7.9. Definition of COMETR sectors

E3ME Sector	NACE Definition
5 Food, drink and tobacco	15, 16
7 Wood and paper	20, 21
10 Pharmaceuticals	24.4
11 Chemicals nes	24 (ex 24.4)
13 Non-metallic mineral products	26
14 Basic metals	27

Source: Cambridge Econometrics.

Table 7.10. Energy as a share of turnover (%)

	DK	DE	NL	FI	SW	UK	SI
Food, drink & tobacco	1.5	2	1.5	1.4	1	1.5	1.9
Wood & paper	1.9	3.3	2.9	5.1	3.7	3	6.5
Pharmaceuticals	0.4	7.2	0	6.5	0.3	0.9	0
Other chems	4.2	6.5	17.5	8.9	8.4	3.9	4.3
Non-metallic minerals	5.4	5.8	4.2	3.5	4.4	4.4	8.9
Basic metals	3	8.7	5.8	6.6	4.5	4.7	9.4

Source: Cambridge Econometrics, E3ME database.

proportions used in the E3ME model. The values are formed by using the coefficients from the base year (2000) input-output tables and are a share of gross output (so wage costs and profit are included in the denominator). These are the sectors that are expected to face the largest increases in energy costs, and therefore face the biggest threat from the ETRs.

We would not expect to see much impact on these sectors in countries where a large proportion of the tax increases fall on households (i.e. The Netherlands and Sweden). It should also be noted that firms in all industries, including energy-intensive ones, will benefit from lower wage costs in countries where there is revenue recycling through employers' social security contributions. Finally, the effects of higher overall growth rates in each country (and the rest of Europe) will give a further boost to product demand.

7.6.5.1 INPUT-OUTPUT ANALYSIS

E3ME's input-output tables for the ETR countries are sourced from Eurostat, except for the UK (ONS) and Slovenia (GTAP database). In each case, they are converted to E3ME's 42 industrial sectors (for chemicals and pharmaceuticals, this usually requires an estimate, as separate data for these sectors are not normally available). The modelling exercise does not use these exact numbers, because input-output coefficients are adjusted on an annual basis following a logistic growth path. However, Table 7.10 still gives a good indication of the importance of energy as an input to each sector and country, with the figures being expressed as a percentage of turnover.

Table 7.10 shows that even in the most energy-intensive industries, energy does not represent a large share of inputs. Only in one case, other chemicals in The Netherlands, does the share of energy inputs in turnover exceed 10 per cent. In most cases, the figure is around 5 per cent, with non-metallic minerals and basic metals apparently having slightly larger shares.

7.6.5.2 PRICE INCREASES

If energy represents around 5 per cent of an industry's input costs (turnover – profit), then even a 50 per cent increase in energy costs is going to lead to only a 2.5 per cent increase in total input costs— even assuming that industry is unable to reduce its fuel consumption or substitute between different fuel inputs. This may or may not be absorbed by firms within the industry (if there were perfect competition within the industry, it would be completely absorbed; if there were no competition, it would be completely passed on). The effect of any price increase will depend on the relevant price elasticities (domestic and export) for the

Table 7.11. Change in industry prices, 2004 (% difference of baseline from reference case)

	DK	DE	NL	FI	SW	UK	SI
Food, drink & tobacco	0.01	0.05	0.00	0.46	1.69	0.00	0.04
Wood & paper	−0.57	−0.40	−0.34	−0.26	−0.33	−0.48	−0.32
Pharmaceuticals	0.01	−0.09	−0.01	0.87	0.05	0.09	−0.02
Other chems	0.32	0.72	0.11	0.36	0.28	0.36	0.08
Non-metallic minerals	0.33	0.46	0.26	0.77	1.06	0.29	0.16
Basic metals	0.51	0.43	0.50	0.53	0.48	0.62	0.46

Source: Cambridge Econometrics.

industry's products. Typically, these would be less than one, so a 2.5 per cent increase in prices would not lead to a 2.5 per cent decrease in product demand. Consequently, even in energy-intensive sectors, we would not expect to see large falls in output.

Table 7.11 shows the results for 2004. This year was chosen because it is the final data point in the input series; by 2004, the ETRs are in place, but there is no blurring of results from the assumption that tax rates remain constant in real terms after 2004.

As expected, the largest increases in prices are in the non-metallic mineral products and basic metals sectors. Prices fall in the wood and paper sector (which operates in an EU market rather than national markets). This is mainly due to a reduction in labour costs in the sector (which form a much larger share of input costs than energy does), and this reduction is mainly a result of reductions in social security payments in Germany and the UK.

Only two of the sectors show price rises above 1 per cent. These are both in Sweden, where the effects are actually an indirect result of higher consumer prices, particularly in electricity (from the ETR), which in turn leads to an increase in wages. In most other cases (excluding wood and paper), the differences are in the range of 0.2–0.4 per cent. In most cases, the price increases also include a factor for an increase in investment. This mainly represents firms' decisions to purchase new machinery in response to higher energy prices. While this may have a negative short-term effect on price competitiveness, it will improve long-term non-price competitiveness through the production of higher-quality output (which may again command higher prices).

Table 7.12 compares the effects of the ETR on export prices in an energy-intensive industry, basic metals, and a non-energy-intensive industry, pharmaceuticals. Although basic metals is an energy-intensive industry,

Table 7.12. Change in export prices, 2004 (% difference of baseline from reference case)

	DK	DE	NL	FI	SW	UK	SI
Basic metals	0.49	0.41	0.50	0.58	0.51	0.90	0.46
Pharmaceuticals	0.20	0.82	0.00	0.22	0.21	0.16	0.00

Source: Cambridge Econometrics.

the effects on export prices are small for two reasons: first, export prices for basic metals are to a large extent decided by world commodity markets and secondly, because a number of energy-intensive industries are exempt from the ETR. However, in all the ETR countries, with the exception of Germany, export prices were higher in 2004 for basic metals than for pharmaceuticals.

7.6.5.3 CHANGES IN OUTPUT

The effects of the ETRs on industry output are less easy to interpret because they include a number of different factors:

- price effects outlined above;
- non-price effects from additional investment;
- consumer demand;
- activity in export markets;
- production in competing import markets.

Table 7.13 shows the percentage increase or decrease in gross output at factor cost (which excludes tax payments) for each of the examined industries, again in 2004. The results show that, in many cases, the overriding effect is higher domestic demand from consumers. In most cases, gross output in the affected industries increases slightly. This is not entirely unexpected, given the modest nature of the price increases recorded. The scale of the increases varies across sectors much more than across countries. The smallest differences are in the UK, where the ETR was smallest. This suggests that domestic demand is a key determinant in industry output.

Food and drink in Sweden is a special case in the results: prices do rise in Sweden in the food and drink industry (see Table 7.11). However, this is not by as much as the overall consumer price index, which rises primarily due to electricity costs. Consequently, food and drink becomes comparatively cheaper and receives a larger share of consumer spending:

Table 7.13. Increase in industry gross output, 2004 (% difference of baseline from reference case)

	DK	DE	NL	FI	SW	UK	SI
Food, drink & tobacco	0.65	0.56	0.13	0.64	4.24	0.02	0.28
Wood & paper	0.29	0.17	−0.27	0.06	0.19	0.04	0.04
Pharmaceuticals	0.08	−0.02	−0.06	0.14	−0.05	0.00	−0.04
Other chems	0.03	0.00	0.00	0.31	0.46	−0.07	−0.07
Non-metallic minerals	0.08	−0.28	0.05	0.54	0.31	−0.03	0.02
Basic metals	0.08	−0.15	0.63	0.08	0.08	−0.16	0.00

Source: Cambridge Econometrics.

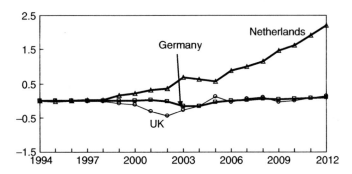

Figure 7.14. The effect of ETR on gross output of basic metals

Note: % difference is the difference between the base case and the counterfactual reference case.

Source: Cambridge Econometrics.

in turn, consumer spending is boosted overall by reductions in income tax. Consumer demand accounts for half of gross output in the food and drink industry.

Figure 7.14 shows that gross output in the basic metals sector remains largely unchanged in most of the ETR countries. However, in The Netherlands it increases by over 2 per cent. This is due to substantial changes in investment in this industry, as shown in Figure 7.15, which in turn boosts its non-price competitiveness.

The effects of increasing investment are crucial in determining the long-term results in several of the sectors examined. Investment (along with R&D spending) determines product quality and non-price competitiveness, and increased investment can more than compensate for moderate price rises.

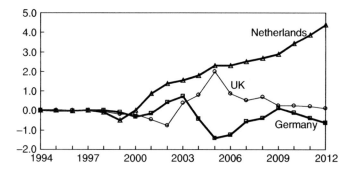

Figure 7.15. The effect of ETR on investment in basic metals

Note: % difference is the difference between the base case and the counterfactual reference case.

Source: Cambridge Econometrics.

One of the effects of the ETR could be that industries invest in lower-carbon technology in order to reduce their overall energy use. However, higher energy prices also represent an element of uncertainty in the economy and may persuade firms to defer or cancel investment plans. Figure 7.15 shows the change in investment in the basic metals sector as a result of the ETR for Germany, the UK, and The Netherlands. It can be clearly seen that the increase in investment in basic metals in The Netherlands due to the ETR results in increased gross output.

7.6.6 The effects of ETR: country results

7.6.6.1 THE ETR IN DENMARK

The Danish ETR raised the effective tax rates of coal and oil by 5–10 per cent, and petrol and electricity by 10–15 per cent. There was little change in the price of natural gas. According to the tax revenue data, taxes were highest in relative terms in the 1990s, and were not increased in line with higher energy prices (electricity taxes for industry are the exception to this), or the introduction of the ETS in 2005. Therefore the largest relative effects from the ETR are near the start of the period, as shown in Figure 7.16.

It is not surprising therefore that the reduction in energy demand is greatest in the 1990s; after this point, the effects of the tax are diluted by higher energy prices (see Figure 7.17). This would appear to illustrate the fact that the Danish government could have achieved a larger decrease in

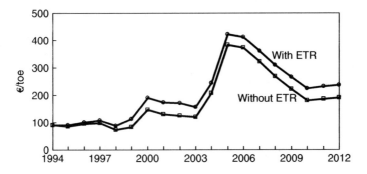

Figure 7.16. Coal prices in Denmark
Source: Cambridge Econometrics.

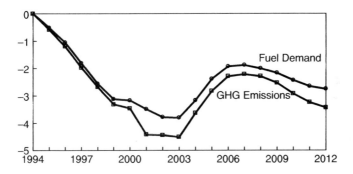

Figure 7.17. The effects of ETR: fuel demand and GHG emissions in Denmark
Note: % difference is the difference between the base case and the counterfactual reference case.
Source: Cambridge Econometrics.

emissions if it had held constant environmental tax as a share of energy prices in the period when energy prices rose.

It should also be noted that, where possible, Danish industry switches to using natural gas as an input rather than coal or electricity, but as power generation dominates demand for gas in Denmark, a fall in electricity consumption leads to a fall in the demand for gas.

In conducting this analysis, one should not ignore the effects of taxes on motor spirit. The ETR leads to a 5 per cent fall in the demand for middle distillates, and this is an important factor explaining the difference between the fall in demand for fuel, and the fall in emissions.

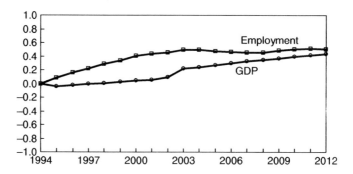

Figure 7.18. The effects of ETR: GDP and employment in Denmark

Note: % difference is the difference between the base case and the counter factual reference case.

Source: Cambridge Econometrics.

7.6.6.2 ECONOMIC EFFECTS OF ETR IN DENMARK

Apart from a small increase in investment, all of the government revenues from ETR are recycled through reductions in employers' social security contributions. This proved to be a very effective way of increasing economic activity, with immediate increases in employment leading to higher average incomes, household consumption, and GDP growth (Figure 7.18 shows the GDP and employment effects). The largest increases in employment came in the retail and construction sectors (these two sectors account for around half of the overall growth in employment).

The ETR had very little impact on international trade into and out of Denmark, with virtually no change in exports and a small increase in imports in line with overall GDP growth.

There are no clear inflationary effects resulting from the Danish ETR, with prices falling in several sectors as a result of lower unit labour costs. This is a key factor in explaining why there is not a relative decline in aggregate exports from Denmark between the baseline and reference cases.

7.6.6.3 THE ETR IN GERMANY

The German ETR started in 1999. It focused mainly on business use of energy, but included a component for household electricity use. Coal was not included in the German tax reforms.

Taking this into account, it is not surprising that there is a wide variation in energy price rises across sectors, depending on the fuel inputs to each one. For example, in power generation, where coal is a major input,

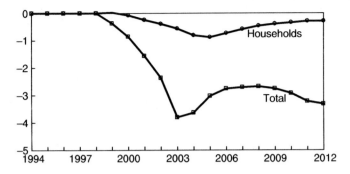

Figure 7.19. The effects of ETR: fuel demand in Germany

Note: % difference is the difference between the base case and the counter factual reference case.

Source: Cambridge Econometrics.

average energy prices only rose by 4 per cent, but in food, drink, and tobacco, the increase was more than 15 per cent. The increase in average road transport costs, and average energy prices for households, was less than 5 per cent.

Overall, the ETR reduced energy demand by around 3 per cent (see Figure 7.19). After 2003, the decrease was less, due to higher energy prices making fuel taxes relatively smaller, but this increased again as energy prices decline in real terms over 2006–10. This partly reflects the assumption that energy taxes increase in line with consumer price inflation, not energy costs, over the forecast period.

Unsurprisingly, given that taxes focus on industry rather than households, the initial reduction in energy demand from households is far smaller and diminishes over time. The only household tax included in the ETR was on electricity.

The predicted fall in GHG emissions in Germany is linked very closely to the fall in total fuel consumption. The main reason that emissions do not fall by more than fuel demand is that the German ETR does not include a tax on coal, so there is in fact a small increase in demand for coal in Germany, as some industries switch to cheaper fuel options. It should be noted that, were it not for the inclusion of the emissions trading scheme, the increase in demand for coal would probably have been higher.

7.6.6.4 ECONOMIC EFFECTS OF ETR IN GERMANY

The ETR in Germany produced a modest increase in GDP, around 0.2 per cent in 2006, increasing to 0.4 per cent over the forecast period to 2012.

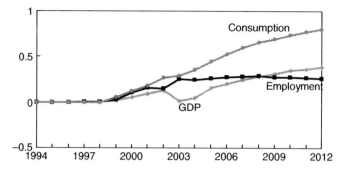

Figure 7.20. The effects of ETR: GDP, consumption and employment in Germany

Note: % difference is the difference between the base case and the counter factual reference case.

Source: Cambridge Econometrics.

There is one notable slowdown in 2003–4 when the tax revenues from gas use increased substantially, as the power generation sector accounts for most gas consumption. The rise in gas use led to a strong increase in electricity prices. This had a negative short-term impact on investment (although it did slightly boost employment). However, once energy prices started to rise and the ETS was introduced in 2005, the effects of the ETR became much smaller in relative terms and growth in investment and GDP became faster.

The combination of reductions in employers' and employees' social security contributions provides a direct boost to income, but also lowers labour costs and so increases employment and therefore average incomes. Figure 7.20 shows that it is in fact consumption that drives GDP growth, more than compensating for a slight worsening in trade performance (although part of the increase in imports will be a result of higher domestic household consumption).

7.6.6.5 THE ETR IN THE NETHERLANDS

There are certain similarities between the Dutch and Swedish tax reforms, in that the bulk of the new revenues come from households' use of electricity. Although households and industry pay the same tax rates in The Netherlands, industry benefits from lower rates for high-volume users (households were assumed to fall into the lowest-volume category). The Dutch ETR was introduced in 1998, but the main effects come through from 1999. Aside from the increases in electricity prices, there was also an increase in taxes on gas (again with lower rates available for high-volume users), but very little change in taxes on coal, heavy oil, and motor spirit.

The effect on the electricity price was greatest in 2001, when the tax reforms almost doubled prices to households. However, revenues fell over 2001–4 and energy prices increased, as Figure 7.21 shows, and so effective tax rates fell as a share of total price, lessening the final impact.

7.6.6.6 ENERGY DEMAND IN THE NETHERLANDS

Although price increases for gas were smaller, consumer demand for gas fell much more sharply than for electricity as a result of the tax reforms (see Figure 7.22). Electricity demand actually rebounded

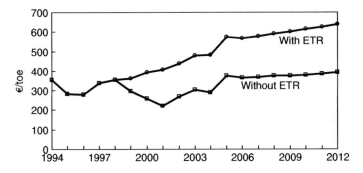

Figure 7.21. The effects of ETR: average household energy prices in The Netherlands

Source: Cambridge Econometrics.

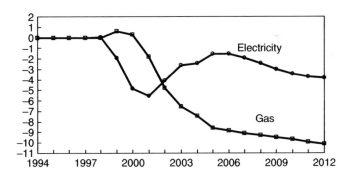

Figure 7.22. The effects of ETR: household energy demand in The Netherlands

Note: % difference is the difference between the base case and the counterfactual reference case.

Source: Cambridge Econometrics.

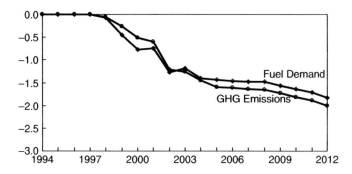

Figure 7.23. The effects of ETR: fuel demand and GHG emissions in The Netherlands

Note: % difference is the difference between the base case and the counterfactual reference case.

Source: Cambridge Econometrics.

somewhat relative to the reference case over the period 2001–5 when electricity taxes fell as a share of the total cost of electricity.

As gas accounts for a much larger share of household energy demand than electricity in The Netherlands, total household demand for fuel falls by more than 8 per cent by 2012.

In other industries (particularly ones that use coal and oil), demand for energy is largely unchanged and increases slightly in some cases. However, this is compensated for by a fall in energy inputs to power generation, reacting to the lower demand for electricity. As 30–40 per cent of Dutch electricity comes from coal, this reduces overall coal demand in The Netherlands, despite there being no direct tax on coal. Because of the combined fall in demand for gas and coal, the difference in overall emissions is fairly similar to the change in aggregate fuel demand (see Figure 7.23).

7.6.6.7 ECONOMIC EFFECTS IN THE NETHERLANDS

The tax revenues from the Dutch reforms were recycled through reductions in income taxes. This boosts overall incomes and household consumption, as Figure 7.24 shows. In contrast, there is almost no change in employment.

GDP does not increase as much as consumer spending because the slightly higher industrial prices have an adverse effect on trade.

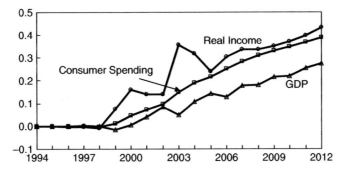

Figure 7.24. The effects of ETR: consumer demand in The Netherlands

Note: % difference is the difference between the base case and the counterfactual reference case.

Source: Cambridge Econometrics.

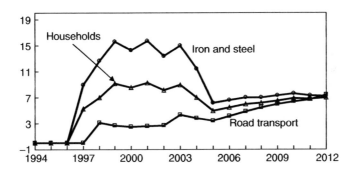

Figure 7.25. The effects of ETR: average fuel prices in Finland

Note: % difference is the difference between the base case and the counterfactual reference case.

Source: Cambridge Econometrics.

7.6.6.8 THE ETR IN FINLAND

Environmental tax reform took place in Finland in 1997. Taxes were increased on all fuels for both business and households. The initial changes increased fuel prices by 10–15 per cent for industry (depending on the fuel mix), and around 9 per cent for households (see Figure 7.25). In the following years, however, higher energy prices diluted the tax effects, particularly in industry dependent on oil and gas. Taxes on road fuels increased in two steps, in 1998 and in 2003, causing an overall increase of around 4 per cent in the cost of motor spirit in 2004. As energy

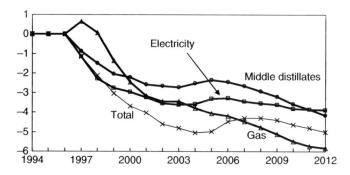

Figure 7.26. The effects of ETR: fuel demand in Finland

Note: % difference is the difference between the base case and the counterfactual reference case.

Source: Cambridge Econometrics.

prices fall over the period 2006–12, the relative price effects of the ETR increase slightly, to around 6 per cent.

The Finnish ETR, as Figure 7.26 shows, reduced fuel demand by around 5 per cent in 2004, and the same amount in 2012. There are falls in demand for all of the main fuels (gas, electricity, and middle distillates) of around 4 per cent in 2004. The demand for coal and heavy oil is reduced even more, but from a lower base. Overall, greenhouse gas emissions fall by about 1 pp more than total fuel use, due to larger reductions in demand for the less widely used (but more polluting) fuels: coal and heavy oil.

7.6.6.9 ECONOMIC EFFECTS IN FINLAND

The effects of higher fuel prices are to reduce real household incomes in Finland and reduce the demand for exports from Finland. Revenue recycling in Finland is not explicit, but is modelled through reductions in direct income taxes, on the grounds that the alternative would have been higher income taxes. This should increase disposable incomes and household consumption, feeding through to the rest of the economy.

However, it turns out that the export effects are not significant, and over the period exports decrease by less than 0.05 per cent. In many sectors, exports increase (the reason is that fuel taxes and revenue recycling will increase investment, which in turn will increase non-price competitiveness).

Finland is the only one of the countries examined to show an increase in GDP even without any revenue recycling. The main reason behind this is that taxes fall almost exclusively on imports of energy products and so, when energy demand falls, there is an improvement in the international trade balance.

In E3ME, lower energy demand is modelled through changing input-output (IO) coefficients. When energy demand falls, the IO coefficients for energy products are reduced. In countries that import most of their energy, this means that imports of energy products fall. In Finland, this is the case for coal and, to a lesser extent, oil. The sector that has the greatest demand for coal and oil is power generation, accounting for 90 per cent of coal demand and 25 per cent of demand for heavy fuel oil. However, electricity is also generated in Finland from biofuels, nuclear power, and renewables. One of the effects of the ETR is to promote these alternative forms of energy, and reduce imports of coal and oil. As electricity prices are assumed to be government regulated, this is assumed to be achieved without a significant loss in real incomes, although company profitability will undoubtedly be affected.

Consumer spending and GDP move very closely together over the historical period, with GDP growth also including reductions in energy imports (see Figure 7.27). Although the difference in GDP growth becomes less in the longer term, there is still a difference of around 0.5 per cent in 2012.

7.6.6.10 THE ETR IN SWEDEN

The data for tax revenues provided by Statistics Sweden are very detailed and provide a disaggregation that can be incorporated directly into E3ME.

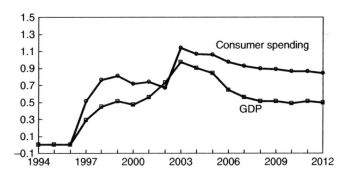

Figure 7.27. The effects of ETR: consumer spending and GDP in Finland

Note: % difference is the difference between the base case and the counterfactual reference case.

Source: Cambridge Econometrics.

The Swedish package of reforms effectively moved much of the tax burden from industry to households, as industry energy taxes are set at zero from 1993 for all the main fuels except motor spirits. This is illustrated in Figure 7.28, which shows the effects of ETR on average fuel prices in the iron and steel industry, and in road transport. In comparison, prices for households (mainly electricity) increased by more than 100 per cent in 2003.

Figure 7.29 separates fuel demand from households from the demand from other sectors, demonstrating the scale of the reduction in demand for energy (mainly electricity) from households.

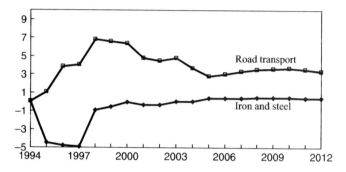

Figure 7.28. The effects of ETR: average fuel prices in Sweden

Note: % difference is the difference between the base case and the counterfactual reference case.
Source: Cambridge Econometrics.

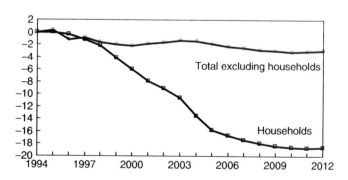

Figure 7.29. The effects of ETR: fuel demand in Sweden

Note: % difference is the difference between the base case and the counterfactual reference case.
Note: Cambridge Econometrics.

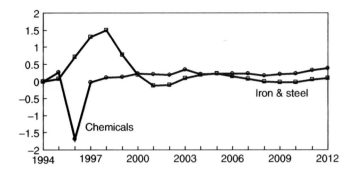

Figure 7.30. The effects of ETR: industrial fuel demand in Sweden

Note: % difference is the difference between the base case and the counterfactual reference case.

Source: Cambridge Econometrics.

In comparison, Figure 7.30 shows that there was little change in demand from most industries. Energy demand in power generation fell as a result of lower household demand for electricity. This illustrates how the ETR was a very effective way of reducing domestic energy demand. A tax increasing in real terms over the period 1994–2004 achieves a reduction in energy demand of 15–20 per cent by 2010.

When considering changes of this scale, it should be noted that estimated price elasticities may not be as accurate as for incremental changes and, as overall household energy demand falls, the price elasticity is likely to become smaller, meaning that the actual fall in demand may be less. The net effect on the rest of the fuel user groups is close to zero.

7.6.6.11 ECONOMIC EFFECTS IN SWEDEN

The effects of increased fuel prices for households and reductions in direct taxes should largely cancel each other out in Sweden, as households are both paying the tax and receiving the benefits. However, the inflationary effects of increased energy prices prove to be a deterrent to short-run consumption, so that, although employment rises very slightly, there is no increase in consumer spending.

One area of the economy that receives an immediate boost is industrial investment. Although this is generally regarded as the most volatile component of GDP, the results show a clear trend that investment increases as a result of higher energy prices. This represents firms' investment in new energy-efficient equipment; and there is an increase in both manufacturing and service sectors.

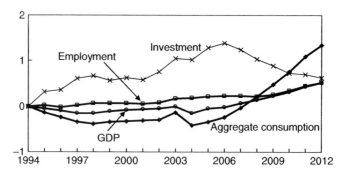

Figure 7.31. The effects of ETR: macroeconomic effects in Sweden

Note: % difference is the difference between the base case and the counterfactual reference case.

Source: Cambridge Econometrics.

Beyond 2006, the tax effects start to fall in relative terms (compared to the reference case), as energy prices rise and the ETS forces costs higher for energy-intensive sectors. The inflationary effects of the ETR are therefore also reduced in relative terms. This, combined with the effects of higher investment over 1994–2005, brings about an increase in consumer spending, which in turn increases GDP and employment. Conversely, investment falls after 2006, suggesting that the boost to consumption will not last much beyond 2010 and in the long run there may be an increase in GDP of something in the range of 0.5 per cent (see Figure 7.31).

7.6.6.12 THE ETR IN THE UNITED KINGDOM

The climate change levy (CCL) in the UK came into effect in 2001, but was announced in 1999 to give companies time to adjust their practices. The aim of the charge was to increase the rate of tax paid by business for fuels used for energy, according to their energy content, to encourage more efficient use of energy and to help the UK meet its GHG targets. Households did not directly pay any of this tax. The CCL is much smaller than some of the other tax reforms examined, raising only €1,200m in 2004. However, in the time between the announcement of the tax and its implementation, the tax gained a lot of media coverage and this raised awareness about the environment in general, particularly in the business sector. The main reason that the announcement effect had such a large impact on commerce was that it was not able to enter into any negotiated agreements, unlike the industrial sectors, and therefore faced a higher tax rate. It should be noted that, in these scenarios, only the price effects (plus the announcement effect in commerce) are being modelled, and any reductions in fuel demand resulting from the negotiated agreements are not included in the difference between the base and reference cases, mainly due to measurement issues. As such, these results are likely to underestimate the full impact of the ETR.

E3ME is not able to model such awareness variables, and to add this feature was outside the remit of COMETR. However, extensive research has been done on the issue with Cambridge Econometrics' UK energy-environment-economy model (MDM-E3) for the UK government (see references), including a separation of price and non-price effects. This research found (through the use of a dummy variable) that there was a substantial reduction in energy use in the retail and commerce sector ('other final use' in E3ME) from the non-price effect, mainly because this sector could not negotiate any reductions in CCL payments. As MDM-E3 and E3ME are very similar in design and structure, it was decided to impose the non-price effects on to the E3ME results exogenously, so that the results presented showed the accumulation of research on the topic.

The study showed that most of the effects of the CCL are attributed to the 'pure' announcement effect, not to the price effect. The effect of the CCL on energy-intensive sectors is far less because most firms in these sectors do not pay the full rate of the Levy, and because no announcement effects are detected in these sectors.

The price increase for each fuel-user group is dependent on whether that group is subject to the charge, and the fuel mix used by that group. In most cases, the increases in energy costs were, as Figure 7.32 shows, small (5% or less). Only sectors with heavy reliance on coal (such as non-metallic mineral products) saw larger increases, and even these differences decreased when world energy prices themselves rose in 2004–5.

Sectors that are more reliant on natural gas, such as food and drink, also faced a slightly higher increase in costs (the data indicate that government revenues from gas and electricity use were roughly equal, despite electricity consumption being around 50 per cent higher). There is no increase in energy costs for households, power generation, or transport sectors. As the price increases in the UK are small and affect sectors that account for only 17 per cent of total energy use, we would not expect to see a particularly large fall in overall energy use and emissions from price effects alone.

In addition to this, fuel demand in the UK tends to be fairly inelastic with respect to price increases. In most of the sectors covered by the CCL, the fall in fuel demand is in the region of 1–2 per cent or less. This is very small when compared to the non-price effects forecast for the commerce

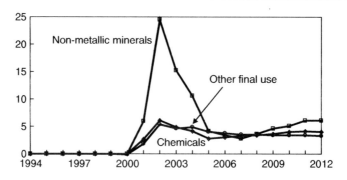

Figure 7.32. The effects of ETR: fuel prices in selected UK sectors

Note: % difference is the difference between the base case and the counterfactual reference case.

Source: Cambridge Econometrics.

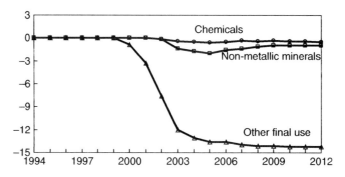

Figure 7.33. The effects of ETR: fuel demand in the UK

Note: % difference is the difference between the base case and the counterfactual reference case.

Source: Cambridge Econometrics.

sector. This sector ('other final use' in E3ME) has, as Figure 7.33 shows, a reduction in fuel demand of 14 per cent by 2012.

7.6.6.13 ECONOMIC EFFECTS IN THE UNITED KINGDOM

Although the UK CCL was able to achieve quite a large reduction in energy demand, this was mostly through non-economic factors, including the announcement effect. In actual fact, the tax levied on British industry was very small and much smaller than the other ETRs in the 1990s. Consequently, we would not expect to see very much change in economic activity. The revenue recycling occurs through reductions in

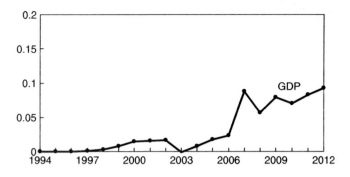

Figure 7.34. The effects of ETR: changes in UK GDP

Note: % difference is the difference between the base case and the counterfactual reference case.

Source: Cambridge Econometrics.

employers' social security contributions, which have the effect of keeping down inflationary factors, but also of raising employment. This leads to a small, but noticeable, effect on GDP over the forecast period, although it should be noted that changes in employment and incomes take several years to produce effects on GDP (see Figure 7.34).

7.6.6.14 THE CO_2 TAX IN SLOVENIA

The CO_2 tax in Slovenia, introduced in 1997, effectively only increased taxation of natural gas over the period 1994–2004, with a higher rate for households. Power generation was exempt from the tax and was therefore not included in the analysis.

Overall demand for gas fell by around 1.5 per cent in 2004. As in many of the other countries, tax rates were not increased in line with higher energy costs. There was very little change in demand for other fuels, with only a small increase in electricity demand as households switch from gas. Therefore, the overall impact on fuel demand was much less than the effect on the demand for gas, at only around 0.3 per cent (see Figure 7.35).

7.6.6.15 FUEL DEMAND IN SLOVENIA

Not surprisingly, as the tax is on natural gas, one of the cleanest fuels, the reduction in greenhouse-gas emissions is much lower than the overall reduction in fuel demand.

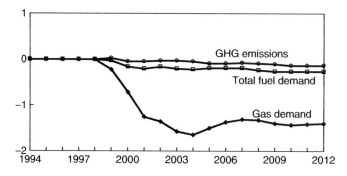

Figure 7.35. The effects of ETR: fuel demand and GHG emissions in Slovenia

Note: % difference is the difference between the base case and the counterfactual reference case.

Source: Cambridge Econometrics.

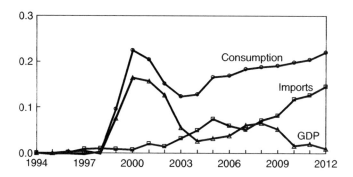

Figure 7.36. The effects of ETR: economic effects in Slovenia

Note: % difference is the difference between the base case and the counterfactual reference case.

Source: Cambridge Econometrics.

7.6.6.16 ECONOMIC EFFECTS IN SLOVENIA

Revenue recycling in Slovenia was assumed to be in the form of reductions in income tax. This gives an immediate boost to real household incomes and consumption. However, there is virtually no resulting increase in employment. In the longer term, imports increase as a result of higher domestic prices in some of the energy-intensive sectors, notably basic metals (which is important to Slovenia). As trade plays a very important role in the Slovenian economy, this reduces the overall effects on GDP.

The immediate effect of the reforms is a small (around 0.15%) increase in GDP. In the long term, there is no increase in GDP attributable to the tax reforms.

As the Slovenian tax was not a true ETR, and had no attached revenue recycling mechanism, it is perhaps more appropriate to consider the case with only the tax, and noting that there is an overall increase in the tax burden. The results for energy demand are largely unchanged from Figure 7.35 above (as energy prices are roughly the same with or without revenue recycling). GDP, however, falls by around 0.05 per cent as a result of the environmental tax with no revenue recycling. This is mainly due to a fall in export volumes of nearly 0.1 per cent.

References

Adams, J. 1997. 'Globalisation, trade, and environment, in OECD, *Globalisation and Environment: Preliminary Perspectives*. OECD proceedings. Paris, OECD, 179–97.

Almon, C. 1991. 'The INFORUM approach to inter-industry modelling'. *Economic Systems Research*, 3/1: 1–7.

Barker, T. 1998. 'The effects of competitiveness of coordinated versus unilateral fiscal policies reducing GHG emissions in the EU: an assessment of a 10% reduction by 2010 using the E3ME model'. *Energy Policy*, 26/14: 1083–98.

——and Köhler, J. 1998. *International Competitiveness and Environmental Policies*. Cheltenham: Edward Elgar.

——and Peterson, W. (eds.) 1987. *The Cambridge Multisectoral Dynamic Model of the British Economy*. Cambridge: Cambridge University Press.

Boltho, A. 1996. 'Assessment: international competitiveness'. *Oxford Review of Economic Policy*, 12: 1–16.

Cambridge Econometrics. 2005. 'Modelling the initial effects of the Climate Change Levy'. A Report Submitted to HM Customs and Excise by Cambridge Econometrics, Department of Applied Economics, University of Cambridge and the Policy Studies Institute.

Grubb, M., Hope, C., and Fouquet, R. 2002a. 'Climatic implications of the Kyoto Protocol: the contribution of international spillover'. *Climatic Change*, 54: 11–28.

——Köhler, J., and Anderson, D. 2002b. 'Induced technical change in energy and environmental modelling: analytical approaches and policy implications'. *Annual Review—Energy Environment*, 27: 271–308.

Kohlhaas, M. 2005. *Gesamtwirtschaftliche Effekte der ökologischen Steuerreform*. Berlin: Deutsches Institut für Wirtschaftsforschung.

Lee, K. C., Pesaran, M. H., and Pierse, R. G. 1990. 'Aggregation bias and labour demand equations for the UK economy', in T. S. Barker and M. H. Pesaran (eds.), *Disaggregation in Economic Modelling*. London: Routledge.

Porter, M. 1991. 'America's green strategy'. *Scientific American*, 264: 168.

——and van der Linde, C. 1995. 'Toward a new conception of the environment-competitiveness relationship'. *Journal of Economic Perspectives*, 9/4: 97–118.

Reinaud, J. 2005. 'Industrial competitiveness under the European Union Emissions Trading Scheme'. IEA Information paper.

Sijm, J. P. M., Kuik, O. J., Patel, M., Oikonomou, V., Worrell, E., Lako, P., Annevelink, E., Nabuurs, G. J., and Elbersen, H. W. 2004. 'Spillovers of climate policy: An assessment of the incidence of carbon leakage and induced technological change due to CO_2 abatement measures'. Netherlands Research Programme on Climate Change Scientific Assessment and Policy Analysis. Energy Research Centre of The Netherlands.

8

Carbon Leakage from Unilateral Environmental Tax Reforms in Europe, 1995–2005[1]

Terry Barker[2], Sudhir Junankar[3], Hector Pollitt[4], and Philip Summerton[5]

8.1 Introduction

Carbon leakage is one measure of the effectiveness of unilateral policies to reduce CO_2 emissions. Carbon leakage is measured by taking the increase in CO_2 emissions outside the country or region taking domestic mitigation action and then dividing by the reduction in the emissions of the country or region. It is an important measure because there are concerns about the effectiveness of unilateral action, either by one country acting alone in an environmental tax reform (ETR), or in the EU acting as a bloc, when there is potential for carbon-intensive production to migrate outside the country or region taking action.

Most of the literature on carbon leakage is about prospective leakage from policies which are being considered or which are just coming into force, nearly always using computable general equilibrium (CGE) models based on one year's data, with a very stylized treatment of the dynamics

[1] This chapter reproduces an article by the authors with the same title, first published in *Energy Policy*, 35, 2007, 6281–92. With permission from Elsevier.
[2] Terry Barker, Director, Cambridge Centre for Climate Change Mitigation Research, Department of Land Economy, University of Cambridge, United Kingdom and Cambridge Econometrics, United Kingdom.
[3] Sudhir Junankar, Manager, Cambridge Econometrics, United Kingdom.
[4] Hector Pollitt, Project Manager, Cambridge Econometrics, United Kingdom.
[5] Philip Summerton, Economist, Cambridge Econometrics, United Kingdom.

of policy effects. Such analysis is *ex ante*, in that it concerns future effects, so that there is no immediate check against actual outcomes. Our study develops the literature, in that it assesses potential leakage from historical actions and uses an econometric dynamic model, that is, it is an *ex post* analysis of actual carbon policies as components of tax reforms. It is also one of the few studies to assess carbon leakage when revenues from policies are explicitly recycled by governments into reducing other taxes or raising government expenditure, rather than being neutralized by being returned (lump-sum) to consumers. We have studied six examples of ETRs in Europe over the period since 1994, using Cambridge Econometrics' E3 model of Europe (E3ME [6]) and developed a series of scenarios to assess the nature and extent of carbon leakage, both short and long term, annually, to the year 2012, with a variety of recycling schemes.

Section 8.2 provides a brief review of the literature on carbon leakage, which is dominated by debates arising from CGE modelling. Section 8.3 describes the approach taken here to modelling the effects. Section 8.4 describes the policies incorporated into this modelling and the scenarios used. Section 8.5 describes the results, including the overall impacts of ETR policies on economic activity and CO_2 emissions, and the sources and magnitude of carbon leakage effects. Section 8.6 provides some conclusions.

8.2 The literature on carbon leakage

The IPCC's *Second Assessment Report (SAR): 1995* (1996) found a high range of variation in leakage rates for OECD action given by world models, going from close to zero to 70 per cent. The *Third Assessment Report* (TAR) (2001) found that the range had narrowed to 5–20 per cent, but noted that these estimates come from models with similar treatment and assumptions. It also noted that the narrower range does not necessarily reflect more widespread agreement. The TAR found that international permit trading substantially reduces leakage. The TAR also considered spillovers through improvement in performance or reduction in cost of low-carbon technologies.

Over the last few years, the literature has extended the analysis, using equilibrium models to include effects of trade liberalization and

[6] For more details, the reader should refer to the model website <http://www.e3me.com> and the online manual <http://www.camecon-e3memanual.com/cgi-bin/EPW_CGI>.

increasing returns in energy-intensive industries; and a new empirical literature has developed.

8.2.1 *Equilibrium modelling of carbon leakage from the 1997 Kyoto proposal*

Gerlagh and Kuik (2006) provide a review of the estimated leakage rates in the CGE literature and a meta-analysis explaining the effects of assumptions in the models on the results. Here we focus on some significant studies.

Paltsev (2001) uses a static global equilibrium model GTAP-EG based on 1995 data to analyse the effects of the 1997 proposed Kyoto Protocol. He reports a leakage rate of 10.5 per cent, within a sensitivity range of 5–15 per cent covering different assumptions about aggregation, trade elasticities, and capital mobility, but his main purpose is to trace back non-Annex B increases in CO_2 to their sources in the regions and sectors of Annex B. The chemicals and iron and steel sectors contribute the most (20% and 16% respectively), with the EU being the largest regional source (41% of total leakage). The highest bilateral leakage is from the EU to China (over 10% of the total).

Kuik and Gerlagh (2003), using the similar GTAP-E model, conclude that for Annex I Kyoto-style action, 'carbon leakage is modest, confirming an extensive set of earlier studies'. They find that the major reason for the leakage is the reduction in world energy prices, rather than substitution within Annex I. They find that the central estimate of 11 per cent leakage is sensitive to assumptions about trade-substitution elasticities and fossil-fuel supply elasticities and to lower import tariffs under the Uruguay Round. These sensitivities result in a range of 6–17 per cent leakage. In a more recent application (Gerlagh and Kuik, 2006), the model is extended to include technological spillovers and the leakage rates are much reduced, even becoming negative under some assumptions.

In contrast to this consensus of global leakage for Kyoto-style action of about 10 per cent, Babiker's (2005) paper presents findings that extend those reported in the SAR and the TAR. He extends a seven-region, seven-good, and three-industry global CGE model (similar to the other GTAP models except for the energy-intensive sector and the earlier 1992 database). The distinctive extension is the inclusion of a treatment of increasing returns to scale and strategic behaviour in the energy-intensive industry. Assuming the adoption of the Kyoto Protocol by the OECD region, he presents four leakage rates, which depend on the assumptions adopted:

- 20 per cent for constant returns to scale and differentiated products (the Armington assumption);
- 25 per cent for increasing returns to scale (IRTS) and differentiated products;
- 60 per cent for constant returns and homogeneous goods (HG);
- 130 per cent for the HG-IRTS combination.

The main reason for the higher estimates is the inclusion of a treatment of increasing returns to scale and strategic behaviour in the energy-intensive industry. The 130 per cent rate implies that OECD action leads to more global GHG emissions rather than less.

8.2.2 Assessing very high rates of carbon leakage

In assessing this high leakage finding, it is important to understand the critical underlying assumptions.

- The CGE model assumes a global social planner to maximize welfare, full information over space and time, perfect competition, and identical firms in each sector ('representative agents').

- The composite energy-intensive good is treated as homogeneous. The high leakage rates come when the composite energy-intensive good has to pay carbon taxes or emission permit prices, and relocates abroad. The implicit assumptions of perfect substitution and no transport costs mean that production relocates without extra cost. However the composite good includes paper and pulp, chemicals, and metals; so it is clearly very mixed in terms of technologies in supply and uses in demand. In fact, one country's production is not perfectly substitutable for that of another as assumed, since the mix will differ.

- Increasing returns are included in only one sector. Adopting this assumption for the energy-intensive industry alone seems arbitrary, since many other products are produced under increasing returns (electricity, machinery, vehicles, computers, software, and communications). Indeed the literature (e.g. McDonald and Schrattenholzer, 2001) does not emphasize the technologies used by energy-intensive industries. In consequence, given perfect substitution, all production is likely to relocate, depending on the assumed dynamics in the model, and with increasing returns, the production in the non-Kyoto countries will become more price competitive, hence the 130 per cent leakage rates.

- Adjustment to a new equilibrium is assumed to take place over many years (e.g. 18 years (1992 to 2010), when the calibrated base year is 1992, with a solution for Kyoto effects for 2010). In fact, Kyoto action has largely taken place after ratification in February 2005. For example, the EU emission trading scheme began in 2005. The result is a much shorter time for leakage than that assumed in the study. The structure of international trade has also changed substantially since the early 1990s, with developing countries, China in particular, becoming much more important in international trade.

Babiker's (2005) model shows that energy-intensive industries will re-locate in response to the change in relative prices brought about by a 28 per cent carbon abatement below business as usual by 2010 (the paper does not state which policy is assumed). The result shows the potential for international trade to undermine unilateral environmental policies under special assumptions and conditions. In fact, mitigation action has tended to give preferential treatment to energy-intensive industries, and any trade quotas, for example, steel quotas, will obstruct relocation.

The policy implications of such findings are that carbon leakage is potentially a serious threat to the effectiveness of mitigation policies. Special treatment of the energy-intensive sectors most affected reduces the threat, but also the overall benefits of the policies. The weakness of the equilibrium modelling is that it is based on one year's data and assumptions such as global maximization of private consumption, homo-geneous goods, constant returns to scale, and perfect competition. The Babiker study shows that including increasing returns to scale in one sector in such models under an assumption of perfect substitution can lead to the wholesale transfer of that sector's output, that is, there are special conditions under which industries will re-locate. However, such extreme results are not found in the empirical studies of carbon leakage as a general response to mitigation under the Kyoto Protocol.

Sijm *et al.* (2004) summarize the reasons that the models can yield such extreme results. 'Models provide a useful, but abstract tool for climate policy analysis; they are faced by several problems and limitations with regard to practical policy decision-making, including problems such as model pre-selection, parameter specification, statistical testing or empiri-cal validation' (p. 14).

8.2.3 *Empirical analysis of carbon leakage from the EU ETS*

Sijm *et al.* (2004) also provide an empirical analysis of carbon leakage from energy-intensive industries. The authors argue that the simple indicator of carbon leakage is insufficient for policy-making. The potential beneficial effect of technology transfer to developing countries arising from technological development brought about by Annex I action is substantial for energy-intensive industries, but has so far not been quantified in a reliable manner. 'Even in a world of pricing CO_2 emissions, there is a good chance that net spillover effects are positive given the unexploited no-regret potentials and the technology and know-how transfer by foreign trade and educational impulses from Annex I countries to Non-Annex I countries' (p. 179).

In the empirical analysis of effects in energy-intensive industries, there are many other factors besides the price competitiveness considered in the modelling studies reporting high leakage rates. They conclude that, in practice, carbon leakage is unlikely to be substantial, because transport costs, local market conditions, product variety, and incomplete information all favour local production. They argue that the simple indicator of carbon leakage is insufficient for policy-making.

Using a detailed model of the world industry, Szabo *et al.* (2006) report production leakage estimates of 29 per cent for cement, with an EU ETS allowance price of €40/tonne CO_2. Demailly and Quirion (2006), using the same model, estimate leakage for cement for '90%-grandfathered allowances' at around 50 per cent, but the rate is very sensitive to assumptions about allocation and auctioning. Leakage rates tend to rise, the higher the allowance price. More generally, Reinaud (2005) surveys estimates of leakage for energy-intensive industries (steel, cement, newsprint, and aluminium) with the EU ETS. She comes to a similar conclusion to Sijm *et al.* (2004) and finds that with the free allocation of CO_2 allowances, 'any leakage would be considerably lower than previously projected, at least in the near term' (p. 10). However, 'the ambiguous results of the empirical studies in both positive and negative spillovers... warrant further research in this field' (p. 179). Analytical studies of climate policy models that focus on the steel industry found that the stricter the climate policy, the higher the rate of carbon leakage. With carbon prices of around €10/tonne CO_2, rates of around 25–40 per cent of carbon leakage from Annex I to non-Annex I countries were found to be due to the relocation of production. Two of the models also found that leakage was greater with increases in tax rates at low carbon prices compared with high prices.

However, there are uncertainties surrounding these models, as they are not specified to consider whether elasticity of demand for products determines the location of production across countries. The models also try to estimate the impact of future, rather than past, climate change policies on the incidence of carbon leakage. There is no representation of technological spillovers from industrialized to developing countries, which are considered the most important market for technology implementation. These technologies reduce the demand for fuel use and therefore the level of CO_2 emissions. Thus, these models have not provided significant evidence that environmental regulation promotes the relocation of high-polluting industries.

Sijm *et al.* (2004) also argued from their empirical analysis that environmental policy has not been a significant decision criterion for the location of investment and that it is not a key explanatory factor for investing in energy-intensive processes in developing countries, as the cost effects of environmental regulation are found to be small. Past experience also suggests that shifts in production shares in the global market have not clearly been due to past environmental policy changes. The production shift has been driven by market size, growth in regional demand (due to developments in new markets and increasing demand in developing countries), and wage levels rather than by a decrease in the competitiveness of industrialized countries compared with developing countries. This has been observed for the steel industry, where strong demand for these products has seen a shift in production to developing countries such as China. Even if relocation in production to developing countries occurs, industries such as iron and steel tend to use the most recent technology, as this minimizes planning costs and maintenance costs. Therefore, it is not obvious that the cost effects of environment policy are influential motives for relocation.

For the purpose of investigating carbon leakage within the EU, there is not enough literature on carbon leakage to warrant conclusions about the effects of the climate change policies of one member state on emissions elsewhere in the EU. The same arguments apply as to those between Annex I and non-Annex I countries in a global context, but technological transfer within the EU is likely to be easier, and the cross-border activities of multinationals are more extensive. Barker (1998) provides estimates of leakage from unilateral policy action by EU member states for a 10 per cent reduction in GHGs by 2010, using additional excise duties on the carbon content of energy, with revenues recycled by reducing employers' social security contributions. These can be found in Table 8.1.

Table 8.1. Projections of CO_2 reductions in EU member states, 2010

	Difference from base case in mtC			Carbon leakage (%)
	Unilateral action internal to the Member State	Effects of this unilateral action in the rest of the EU	Total effects of action internal to the Member State	
Belgium	−5.8	0.1	−5.7	1.8
Denmark	−0.9	0.0	−0.9	0.0
West Germany	−16.9	0.1	−16.8	0.6
Spain	−8.0	−0.6	−8.6	−7.0
France	−14.7	−0.6	−15.3	−3.9
Ireland	−1.3	0.0	−1.3	0.0
Italy	−13.1	−0.1	−13.2	−0.8
Luxembourg	−0.2	0.0	−0.2	0.0
Netherlands	−2.5	0.1	−2.4	4.2
Portugal	−1.0	0.0	−1.0	0.0
United Kingdom	−14.9	0.3	−14.6	2.1

Source: Adapted from Barker (1998: p. 1094).

The results show that leakage can raise or lower emissions, but the estimates of leakage are very small in relation to the effects in the countries taking action. The negative leakage for France and Spain comes from unilateral actions leading to an improvement in competitiveness, a fall in imports from other member states, and a reduction in activity and CO_2 emissions outside the countries. No total is provided, because it is not valid to aggregate the unilateral effects. Note one limitation of this study (and of the results presented below): these leakage rates do not include any leakage outside the EU. Gerlagh and Kuik (2006) calculate (using CGE modelling) that the effects of unilateral action on world energy prices can be substantial. However, the impact of the ETRs on European oil imports from the rest of the world is likely to be very small, if not negligible, since the ETRs are not generally focused on the transport sector, and they are very modest in scale.

8.3 Modelling carbon leakage

Carbon leakage is measured by the increase in CO_2 emissions outside the countries taking domestic mitigation action divided by the absolute reduction in the emissions of these countries.

$$CL = -\frac{\Delta CO2_N}{\Delta CO2_M} \tag{8.1}$$

where $\Delta CO2_M$ is the change in CO_2 in countries taking mitigating action and $\Delta CO2_N$ is the change in CO_2 in countries not taking mitigating action.

In E3ME modelling, $\Delta CO2_M$ is calculated by subtracting a baseline figure for CO_2 from a counterfactual reference case (discussed below) for the six ETR regions in total. Similarly, $\Delta CO2_N$ is calculated by subtracting the baseline figure for CO_2 from a counterfactual case for any country or group of countries, where mitigation policies (ETR) were not pursued. Carbon leakage is a 'negative spillover' effect and may occur through international trade in energy goods, international trade in other goods and services, international trade in factors of production, or international interaction between government policies. In Sijm *et al.* (2004), there is an extensive discussion of carbon leakage in the context of an increase in CO_2 emissions in non-abating countries due to the implementation of climate policies in Annex I countries. The objective is to determine whether environmental taxes cause carbon leakage through the spatial relocation of production of energy-intensive goods to countries that have not implemented environmental tax reforms (ETRs). This should be observable through patterns in emissions, energy demand, and in international trade.

8.3.1 Methodology of carbon leakage in E3ME

E3ME is large-scale econometric model with a dynamic structure, which is both sector and region specific, that has been designed to model energy-environment-economy interactions between EU member states. It involves the use of econometric estimation to identify the effects of carbon and energy taxes on energy prices, energy demand, fuel use, CO_2 and other emissions, and embed these in a large post-Keynesian non-linear simulation model. The model has been developed in the traditions of the Cambridge dynamic model of the UK economy (Barker and Peterson, 1987), which was further developed to be a regional, European model, E3ME (Barker, 1999; Barker *et al.*, 2001). The effects of technological change modelled this way may turn out to be sufficiently large in a closed global model to account for a substantial proportion of the long-run growth of the system and improvements in energy efficiency over time.

The model is estimated as a set of 22 cross-section time-series equations using cointegration techniques proposed originally by Engle and Granger (1987) and discussed by Abadir (2004) as appropriate for modelling of non-clearing markets in which a long-run solution is not necessarily in equilibrium. E3ME requires as inputs dynamic profiles of population, energy supplies, baseline GDP, government expenditures, tax, and interest and exchange rates; and it derives outputs of carbon dioxide and other greenhouse gas emissions, SO_2 emissions, energy use, and GDP and its expenditure and industrial components.

The version of E3ME used for the analysis is 4.1, which includes the EU25 (as of 2006) individually, 42 industry sectors (including 16 service sectors and a disaggregation of the energy sectors), 28 categories of household spending, 12 fuel types, and 19 distinct fuel-user groups.

Leakage is estimated by developing a set of counterfactual scenarios using E3ME. The specific scenarios are the 'reference case', which is a counterfactual projection without the ETR and the 'baseline case', which is also an endogenous solution of E3ME over the period 1995–2012. The baseline scenario includes the ETR in each member state to be covered by the project, exemptions or special treatment for the industries most affected, and the compensating reduction in another tax (revenue recycling). The difference between the baseline case and the reference case thus gives a dynamic estimate of the impact of ETR policies on the European economies, and enables the calculation of leakage rates.

8.3.2 *Using E3ME to investigate the extent of carbon leakage for the EU*

To determine the extent of any possible carbon leakage, we must consider trade effects and technical change as well as carbon emissions. Environmental tax reforms in E3ME will flow through to exports and imports in countries with and without ETRs (and therefore to output) and to carbon emissions through the following mechanisms.

International trade between countries that have implemented ETRs, and those countries that have not, will be affected by cost and price increases:

- Domestic input costs will increase in the countries with ETRs as a result of higher fuel costs and/or other environmental taxes. Higher input costs raise domestic prices for the products in energy-intensive sectors (but this is also dependent on the extent to which producers pass through the cost increases) relative to those produced in countries without ETRs. If domestic prices rise and the prices are set by the domestic market, this implies that the export prices of products also increase.

- As domestic prices increase, import prices become relatively cheaper for energy-intensive products (long-run price homogeneity[7] is assumed in E3ME).

- Higher export prices imply that other countries without ETRs become more competitive through relatively lower input costs (as there are no energy tax increases). This gives these countries a comparative advantage in energy-intensive products. This may result in the relocation of these industries to countries with less stringent climate-change policies, and we would expect exports to rise from these countries because of relatively lower prices of energy-intensive goods and to meet 'rest of the world' demand. These countries also reduce their imports from countries with ETRs as a result of price effects.

Increased international trade may lead to higher demand for fuel inputs, and thus, CO_2 emissions as a result of the following effect:

- Gross output is higher in countries without ETRs due to lower imports and higher exports but gross output is lower in countries with ETRs.
- Higher gross output from energy-intensive sectors in countries without ETRs may lead to higher energy demand (in production) and emissions. In contrast, lower gross output and energy demand will lead to a reduction in emissions in non-ETR countries.

Figure 8.1 illustrates how the increases in energy and carbon taxes as components of ETR, working through these mechanisms, may lead to carbon leakage in E3ME. However, competitiveness of energy-intensive goods may improve in countries with ETRs, if the policies induce innovation and reduce the energy-intensity of the associated industries. This innovation investment may impact carbon emissions in countries with and without ETRs through technological changes (not illustrated in Figure 8.1). We expect technological change to improve the quality of goods (particularly energy-intensive goods), increasing demand for these goods in domestic and international markets. The effects are introduced into trade, price, and employment equations in the model by a 'technological progress indicator' formed by accumulating past gross investment and R&D expenditures (Barker, 1999). The net impact that improvements in technological change have on CO_2 emissions in countries with and without ETRs depends on:

- the level of increase in investment;
- the willingness of customers to pay for the improved quality of goods;

[7] In E3ME, all price effects are assumed to be relative in the long run.

225

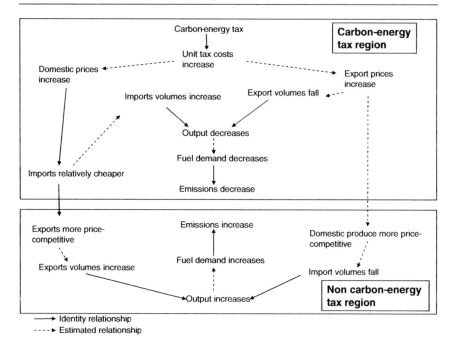

Figure 8.1. How carbon leakage would occur in E3ME

- increases in output due to higher net export demand for improved quality of goods arising from technological change incorporated in the new investment.

In addition, Figure 8.1 only considers the tax effects on unit costs from the carbon/energy tax components of ETR and does not consider revenue recycling. Revenue recycling could have two different effects on the system illustrated in Figure 8.1. First, revenue recycling could cause unit prices to fall in some sectors, in the case where revenues are used to reduce employers' social security contributions, as labour unit costs would fall. However, the effects of this type of recycling will differ between industries and depend on the relative proportions of labour and energy in determining unit cost. For example, the services sector is likely to gain most from the inherent reduction in labour costs, as the services sector is labour intensive. However, the services sector has a low ratio of exports to output, so any offsets on carbon leakage are likely to be small.

Secondly, revenues from ETR may be used to reduce income tax; in this case, output in the domestic industry could rise because of stronger consumer demand. In this second case, carbon leakage might not occur because industries have no incentive to relocate; alternatively, it may cause import demand to increase, as prices become relatively cheaper abroad, and therefore give rise to carbon leakage.

8.3.3 Limitations of E3ME in analysing carbon leakage

E3ME is *not* a world model, and as such the estimates only cover possible carbon leakage to other EU member states that have not implemented ETRs, vis-à-vis those that have, and not to the rest of the world. Estimates will be the leakages via price effects of ETR on energy-intensive industries raising costs, so that non-ETR countries have greater price competitiveness in EU markets than those countries with ETR. Results will include the effects of ETR improving the non-price competitiveness of energy-intensive industries—higher investment leading to more exports and reverse leakage. This is an important finding in empirical studies. Although the volumes for intra- and extra-EU trade can be identified, the same is not true for prices, and therefore the export and import price specifications in the model are for all imports and exports, regardless of destination or source.

8.4 Description of ETR policies and carbon leakage scenarios

The notion of 'environmental tax reform' (ETR) (Ekins and Speck, 2000; Speck, 2006) typically involves the modification of the national tax system to move the burden of taxes from conventional taxes, for example those imposed on labour and capital, to environmentally related activities, such as taxes levied on resource use, especially energy use, or environmental pollution. The implementation of a revenue-neutrality policy is designed to ensure that the tax burden falls more on 'bads' rather than on 'goods', by ensuring that price signals, as a result of the introduction of ETR, give an incentive to households and industries to alter behaviour. Many of these ETRs have involved the introduction of CO_2 taxes (Andersen, 2004). The effects of the tax reforms are complicated by interaction with the effects of other taxes (Bohm, 1997; De Mooij, 2000; Ekins and Barker, 2001; Kratena, 2002). The welfare effects of a prospective ETR for some EU countries are estimated by Barker and Köhler (1998b).

We now describe the ETRs incorporated in this modelling and the scenarios used. Note that it is inevitable that the rich institutional detail of the legislation and the data has to be simplified, stylized, and aggregated in order to undertake a quantitative modelling exercise.

8.4.1 ETRs in six EU MSs

Speck (2006) describes in detail the various ETRs introduced into E3ME and considered in this study. These details have been translated into assumptions, rates of tax, and special treatments in E3ME. The characteristics of the ETRs are listed in Table 8.2, which also includes estimates of revenues from ETRs by country in 2004. ETR as a percentage of GDP in 2004 is less than 1.1 per cent for all the ETR countries and there are noticeable differences between both the tax rates in each country and the time when they were introduced.

Denmark was one of the first countries to implement an ETR in 1992, for households and 1993 for industry. The political objective of the initial reform was to reduce income taxes, but in the second phase of the ETR, this shifted to a reduction in social security contributions, as industry as opposed to households became the focus of the ETR.

Finland was the first country to implement a carbon-energy tax in 1990, which was originally levied on all energy products, with the exception of transport fuels. The first full phase of a Finnish ETR, however, began in 1997. The aim of the tax shift was not revenue neutrality but rather to reduce income taxes; employers' social security contributions were also offset. The Finnish ETR affects both households and industry, but the recycling measures favour households.

Table 8.2. Carbon-energy components of environmental tax reforms in six EU member states, 1995–2005

Dates of introduction	Denmark 1992	Germany 1999	Netherlands 1996	Finland 1997	Sweden 1991	UK 1996
Fuels covered	Coal, oil, electricity and transport fuels	Electricity, gas and transport fuels	Electricity, gas	All fuels	All fuels	Coal, electricity, gas, transport fuels
Sectors affected	Industry and transport	Industry and transport (plus household electricity)	Households	All sectors	Households	Industry and transport
Revenue recycling	Yes	Yes	Yes	Yes	Yes	Yes
Gross revenue, 2004, €m	2,140	18,547	2,287	894	2,585	1,200
GDP, 2004, €m	197,222	2,207,200	489,854	151,935	281,124	1,733,603
ETR as a % GDP	1.08	0.84	0.47	0.59	0.92	0.07

Source: Speck (2006) and Cambridge Econometrics, E3ME database.

The German ETR (see Bach *et al.*, 2002) was implemented in two phases; in the initial stage (1999–2003), various levies were introduced on each major fuel type. The ETR was designed to be revenue neutral, and revenues were recycled through social security contributions. The German government extended the ETR in phase 2 (2004 onwards) by adapting the heating fuels taxes on natural gas and heavy fuel.

In The Netherlands, a carbon-energy tax was added to the tax base in 1990, but in 1996 an ETR was fully implemented with revenue neutrality. All revenues were recycled back to households and industry, but in different ways. For households, this was primarily through a reduction in basic income tax and, for industry, a reduction in the wage component paid by employers and the corporate tax rate.

Sweden introduced the first major ETR in Europe, commencing in 1991, with the objective of reducing personal income taxes. However, in the second stage of the Swedish ETR, social security contributions were also reduced. The ETR in Sweden affected both households and industry, but most of the revenue was recycled to households, particularly in the first stage of the reform.

In the UK, the first sign of an ETR was the introduction of the landfill tax in 1996; this was then followed by the climate change levy in 2001, and subsequently the aggregates tax in 2002. The UK ETR was the smallest ETR (relative to GDP) accounting for just 0.07 per cent. The UK ETR has been revenue neutral and revenues have been recycled through reducing social security contributions. The tax burden also targeted industry and not households.

8.4.2 *The Scenarios*

Leakage is estimated by developing a set of counterfactual scenarios using E3ME. The specific scenarios are as follows.

The reference case is constructed to establish a counterfactual history of the European economy for the period 1995–2012, without the impact of the ETRs implemented over this period. It is a fully dynamic solution of E3ME over the period, given the year-by-year profile of exogenous variables such as other countries' output and prices, exchange rates, interest rates, and fiscal policies in general. It includes the impact of climate and energy policy measures which are not explicitly included in the ETRs and of course the substantial rise in energy prices after 2003.

The baseline case is an alternative fully dynamic solution over the period 1995–2012. This scenario includes the ETR in each member state to be covered by the project, exemptions, or special treatment for the industries most affected and the compensating reduction in another tax (revenue recycling). This scenario is calibrated closely to the observed outcome through using historical data which include the effects of ETR implementation (the historical part of the baseline). The calibration factors are also included in the reference case, so that the two scenarios are comparable. There are substantial differences between ETRs in the use of the revenues from taxes, and this affects the extent of carbon leakage. For example, if most revenues are used to improve the non-price competitiveness of energy-intensive industries, carbon leakage in the long term is likely to be very low.

8.5 Results

8.5.1 Direct analysis of carbon leakage in E3ME

In terms of total carbon leakage to non-ETR member states, the analysis undertaken has assessed the extent of carbon leakage as a result of the ETRs of the six countries collectively and not individually. In other words, the base case includes all the ETRs together and compares this scenario with a reference case without any ETR. Hence these results show the individual and collective changes in carbon emissions resulting from the collective tax reforms of the seven ETR countries.

Figure 8.2 shows the total carbon leakage in non-ETR countries as a result of ETR in the ETR countries considered (Denmark, Germany,

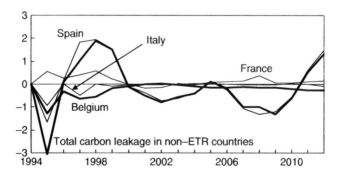

Figure 8.2. Total carbon leakage in non-ETR countries as a result of ETR Carbon Leakage (%)

Note: Carbon leakage is the change in carbon emissions in countries without ETR divided by the change in carbon emissions in ETR countries and expressed as a percentage.

Source: Cambridge Econometrics.

Finland, The Netherlands, Sweden, and the UK). Our results show that carbon leakage is very small in the non-ETR countries as a whole. Carbon leakage fell to −3 per cent in 1995 as a result of the ETR, but from that point on there was no significant carbon leakage (positive or negative) when comparing the baseline case to the counterfactual reference case.

The results also show 'negative' carbon leakage in some years, suggesting a reduction in aggregated carbon emissions in both ETR countries and non-ETR countries. This result is consistent with technological spillover effects. Whereas theory in which most effects are price-based might suggest that an increase in fuel prices through an ETR would cause energy-intensive industries to relocate to countries which have not imposed ETR, and hence lead to carbon leakage, it may in fact be the case that industry invests in energy-efficient processes and technologies, which are then exported to similar industries in the non-ETR regions. There are several other reasons why carbon leakage may not occur to the extent suggested in the previous literature: the cost of relocation, in terms of both transport costs and the costs of finding new markets, may not outweigh the cost of investing in more efficient energy processes. These costs are implicit in the responses of energy demand to relative prices and activities estimated by time-series equations and illustrated by small short-term price elasticities.

Figure 8.2 also provides evidence that carbon leakage in some of the largest non-ETR regions individually was very small when comparing the baseline case with the reference case. This figure shows that in some of the countries where relocation of energy-intensive industries was most likely, namely, France, Spain, Italy, and Belgium, carbon leakage varied between positive and negative over the period. The percentage of carbon leakage was, however, very small in all of these countries, varying between 2 and −2 per cent. As discussed earlier, negative carbon leakage can be explained by technological spillover within Europe and the transfer of new technologies. After the initial negative carbon leakage in 1995, carbon leakage then fluctuates around zero. In Spain, a similar pattern is observed; however, the fluctuations around zero are slightly larger.

Figure 8.3 shows the relative reduction in CO_2 emissions when comparing the baseline case with the reference case. As expected, CO_2 emissions fall in the ETR countries collectively over the period by 3–4 per cent in 2012 as a result of the ETRs. In contrast, the ETRs have almost no effect on the level of CO_2 emissions in non-ETR countries. This suggests that there was no carbon leakage from ETR regions collectively to non-ETR regions.

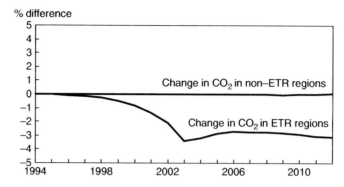

Figure 8.3. Changes in CO_2 emissions in ETR and non-ETR regions % difference

Note: % difference represents the difference between the baseline case and the reference case.
Source: Cambridge Econometrics.

8.5.2 Indirect analysis of carbon leakage in E3ME

Trade patterns will indicate whether carbon leakage has occurred by comparing countries (and sectors) in terms of:

- relative prices of energy-intensive products between countries (reflecting changes in comparative advantage);
- terms of trade (exports/imports) changes as a result of changes in relative prices;
- the ratio between the domestic price (which reflects whether increases in input costs are completely passed through to the end user) and the export price of a product in the country introducing the ETR;
- output changes arising from changes in trade patterns.

Technical change will also indicate whether tax reforms have given incentives to promote investment in more energy-efficient technology. Technical changes may lead to higher quality products being exported from the country introducing the ETR and this may lead to import demand for these products in other countries, even if the prices of imports from countries with ETRs in place are higher. On the other hand, CO_2 emissions may decline in both types of countries if technical improvements occur through positive spillover effects (via foreign direct investment) from the countries developing these technologies.

Figure 8.4 shows the effects of ETR on the exports and imports of the two largest economies considered, Germany and the UK. This figure clearly illustrates that the ETR had very little effect on total intra-EU

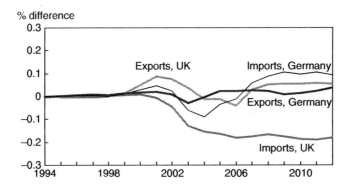

Figure 8.4. Evidence of carbon leakage in exports and imports in the UK and Germany % difference

Note: % difference represents the difference between the baseline case and the reference case.
Source: Cambridge Econometrics.

exports and imports in either Germany or the UK; the difference is between 0.3 and −0.3 per cent, suggesting that it is highly unlikely that carbon leakage would occur, given that there has been little effect on the terms of trade. If carbon leakage were taking place, a fall in exports in both Germany and the UK would result, as firms reinvested in non-ETR regions such as France, Spain, and Italy or in central Europe. In addition, imports to ETR countries would increase if carbon leakage were taking place, as imports would be relatively cheaper. This happens to a very small degree in Germany, but the opposite is true in the UK. This is further evidence to suggest that carbon leakage does not occur as a result of unilateral[8] action regarding ETRs, and in fact goes slightly further to suggest that negative leakage occurs, albeit by a small amount.

At the sectoral level, the results are broadly consistent with the macro-economic results. There is evidence for very small amounts of carbon leakage, but in some cases this is negative. In Sweden, for example, exports of wood and paper are forecast to be 1.1 per cent lower by 2012,[9] as a result of the ETR, suggesting weak evidence for carbon leakage. On the other hand, the basic metals sector in The Netherlands is forecast to see an increase in exports of 2.1 per cent in 2012 when comparing the baseline case with the reference case, suggesting efficiency improvements and investment. This appears to provide evidence contrary to part of the

---

[8] Unilateral, in this case, defines the ETR countries as having taken unilateral action when compared to the rest of Europe which did not undertake ETR.

[9] See Chapter 7.

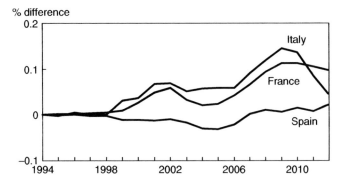

Figure 8.5. Evidence of technological spillover: investment in Spain, France, and Italy % difference

Note: % difference represents the difference the baseline case and the reference case.
Source: Cambridge Econometrics.

carbon leakage argument which suggests that exports will fall, as firms will relocate to countries that have not imposed ETR, and where energy costs are therefore lower.

Furthermore, analysis showed that the ETRs had minimal effects on non-ETR countries. Further disaggregation of differences between the baseline and reference cases highlighted the fact that ETRs have very little effect in non-ETR countries. Figure 8.5 shows that investment and induced technological improvements[10] were minimal in non-ETR countries. Investment as a whole in France, Spain, and Italy changed by less than 0.3 per cent as a result of the ETRs. Whilst this result is small, it may well account for the small changes in CO_2 emissions in France and Italy. Investment in Spain remains virtually constant and does not provide evidence of fluctuations in carbon leakage in Spain as a result of ETR.

8.5.3 Comparing the effects on GDP and GHG in all EU25 countries

Carbon leakage theory suggests that if carbon leakage is to take place, it will happen through changes to international trade in energy goods, international trade in other goods and services, international trade in factors of production, or international interaction between government policies. In terms of our analysis, this implies that if carbon leakage

---

[10] Expenditure in research and development was unchanged between the baseline case and the reference case in all regions.

Table 8.3. The effects of ETR: GDP in EU25 (% difference between the baseline case and the counterfactual reference case)

	2000	2004	2008	2012
Countries with Environmental Tax Reforms (ETR)				
Denmark	0.05	0.24	0.34	0.43
Germany	0.05	0.05	0.29	0.39
Netherlands	0.00	0.11	0.18	0.27
Finland	0.47	0.91	0.51	0.50
Sweden	−0.08	−0.15	0.14	0.52
United Kingdom	0.02	0.01	0.06	0.09
Selected non-ETR EU15				
Austria	0.02	0.03	0.07	0.06
Belgium	0.02	0.02	−0.05	−0.08
France	0.01	0.01	0.05	0.03
Italy	0.00	0.02	0.06	0.06
Portugal	0.00	−0.02	−0.01	0.02
Spain	0.01	0.02	0.03	0.05
Large EU Accession States				
Czech Republic	0.01	−0.01	0.00	0.00
Hungary	0.01	0.01	0.01	0.01
Poland	0.01	0.02	0.02	0.03
Slovenia	0.16	0.03	0.07	0.01
Slovakia	0.02	0.01	0.02	0.00
Total EU 25	0.02	0.04	0.12	0.16

Source: Cambridge Econometrics, E3ME4.1. Ref: flcom.c52.

is taking place, there will be a reduction in GDP for ETR countries when comparing the baseline case with the counterfactual reference case.

Table 8.3 clearly shows that GDP has increased in all of the ETR countries, albeit by a modest amount over the period modelled; this suggests that carbon leakage has not taken place. However, it is possible that due to the ETR the structure of the economy has shifted from energy-intensive industry to non-energy-intensive industry, and GDP has simply increased as a result of revenue recycling. In this case, it is still entirely possible that energy-intensive industries have relocated to non-ETR countries and hence carbon leakage can be said to have occurred.

However, Table 8.4 shows how little greenhouse gas emissions vary as a result of the ETR in the EU25 countries. For the EU as a whole, GDP increases by 0.16 per cent by 2012, whilst greenhouse gases are 1.3 per cent lower, when comparing the baseline to the reference case. This suggests that very little carbon leakage has occurred. France and Spain are the only regions where greenhouse gas emissions increase as a result of ETR action in the ETR countries. There is therefore carbon leakage from

Table 8.4. The effects of ETR: GHG in EU25 (% difference between the baseline case and the counterfactual reference case)

	2000	2004	2008	2012
Countries with Environmental Tax Reforms (ETR)				
Denmark	−3.46	−3.63	−2.30	−3.43
Germany	−0.69	−3.73	−2.68	−3.45
Netherlands	−0.52	−1.46	−1.65	−2.01
Finland	−3.98	−5.90	−4.34	−5.23
Sweden	−3.80	−3.47	−6.35	−6.63
United Kingdom	−0.12	−2.02	−2.42	−1.97
Selected non-ETR EU15				
Austria	0.00	0.02	0.05	0.05
Belgium	−0.01	−0.05	−0.08	−0.13
France	0.00	−0.05	−0.11	0.19
Italy	0.00	0.00	0.00	−0.01
Portugal	0.01	0.00	0.01	0.01
Spain	0.00	0.01	0.08	0.03
Large EU Accession States				
Czech Republic	0.00	0.01	0.01	−0.02
Hungary	0.00	0.00	0.00	0.00
Poland	0.00	0.00	0.00	0.00
Slovenia	−0.05	−0.05	−0.10	−0.13
Slovakia	0.00	0.00	0.00	0.00
Total EU 25	−0.34	−1.47	−1.15	−1.29

Source: Cambridge Econometrics, E3ME4.1. Ref: flcom.c52.

the ETR countries to France and Spain, but not to any other EU25 country by 2012. However, the carbon leakage suggested in 2012 is very small, when compared with previous studies using static CGE modelling, which suggest that carbon leakage might be as high as 20 per cent.

8.6 Conclusions

This study of the potential carbon leakage effects of ETR in six EU member states, which took place over the period 1995–2005, suggests that carbon leakage is not taking place, even at the sectoral level. Output does not appear to be relocating away from ETR countries as a result of the tax switch. Only in a highly competitive, export-driven market does the small industry price increase lead to a decrease in output, namely the UK and German basic metals industries. The absence of strong evidence for carbon leakage is most likely due to the fact that the ETR energy taxes are relatively small and so they do not have a sufficiently large enough effect on unit costs to justify the cost of relocation.

References

Abadir, K. M. 2004. 'Cointegration theory, equilibrium and disequilibrium economics'. *The Manchester School*, 72/1: 60–71.

Andersen, M. S. 2004. 'Vikings and virtues: a decade of CO_2 taxation'. *Climate Policy*, 4/1: 13–24.

Babiker, H. 2005. 'Climate change policy, market structure, and carbon leakage'. *Journal of International Economics*, 65: 421–45.

Bach, S., Kohlhaas, M., Meyer, B., Praetorius, B., and Welsch, H. 2002. 'The effects of environmental fiscal reform in Germany: a simulation study'. *Energy Policy*, 30: 803–11.

Barker, T. 1998. 'The effects of competitiveness of coordinated versus unilateral fiscal policies reducing GHG emissions in the EU: an assessment of a 10% reduction by 2010 using the E3ME model'. *Energy Policy*, 26/14: 1083–98.

—— 1999. 'Achieving a 10% cut in Europe's carbon dioxide emissions using additional excise duties: coordinated, uncoordinated and unilateral action using the econometric model E3ME'. *Economic Systems Research*, 11/4: 401–21.

—— Fingleton, B., Homenidou, B. K., and Lewney, R. 2001. 'The regional Cambridge multisectoral dynamic model of the UK economy', in G. Clarke and M. Madden (eds.), *Regional Science in Business*. Heidelberg: Springer-Verlag, 79–96.

—— and Peterson, W. 1987. *The Cambridge Multi-sectoral Model of the British Economy*. Cambridge: Cambridge University Press.

—— and Köhler, J. 1998a. *International Competitiveness and Environmental Policies*. Cheltenham: Edward Elgar.

—— —— 1998b. 'Equity and ecotax reform in the EU: achieving a 10% reduction in CO_2 emissions using excise duties'. *Fiscal Studies*, 19/4: 375–402.

Bohm, P. 1997. 'Environmental taxation and the double dividend: fact or fallacy', in T. O'Riordan (ed.), *Ecotaxation*. London: Earthscan.

Demailly, D., and Quirion, P. 2006. 'CO_2 abatement, competitiveness and leakage in the European cement industry under the EU ETS: grandfathering versus output-based allocation'. *Climate Policy*, 6: 93–113.

De Mooij, R. A. 2000. *Environmental Taxation and the Double Dividend: Contributions to Economic Analysis*. Amsterdam: North Holland.

Ekins, P., and Barker, T. 2001. 'Carbon taxes and carbon emissions trading'. *Journal of Economic Surveys*, Special Issue, 15/3: 325–76. Also published in N. Hanley and C. Roberts (eds.) 2002. *Issues in Environmental Economics*. Oxford: Blackwell, 75–126.

—— and Speck, S. 2000. 'Proposals of environmental fiscal reforms and the obstacles to their implementation'. *Journal of Environmental Policy & Planning*, 2/2: 93–114.

Engle, R. F., and Granger, C. W. J. 1987. 'Cointegration and error correction: Representation, estimation and testing'. *Econometrica*, 55/2: 251–76.

Gerlagh, R., and Kuik, O. 2006. 'Carbon leakage with international technology spillovers'. Working Paper draft. November.

Kratena, K. 2002. *Environmental Tax Reform and The Labour Market: The Double Dividend in Different Labour Market Regimes*. Cheltenham: Edward Elgar.

Kuik, O., and Gerlagh, R. 2003. 'Trade liberalization and carbon leakage'. *Energy Journal*, 24: 97–120.

McDonald, A., and Schrattenholzer, L., 2001. 'Learning rates for energy technologies'. *Energy Policy*, 29/4: 255–61.

Paltsev, S. 2001. 'The Kyoto Protocol: regional and sectoral contributions to the carbon leakage'. *Energy Journal*, 22: 53–79.

Sijm, J. P. M., Kuik, O. J., Patel, M., Oikonomou, V., Worrell, E., Lako, P., Annevelink, E., Nabuurs, G. J., and Elbersen, H. W. 2004. 'Spillovers of climate policy: an assessment of the incidence of carbon leakage and induced technological change due to CO_2 abatement measures'. Netherlands Research Programme on Climate Change Scientific Assessment and Policy Analysis. Energy Research Centre of The Netherlands.

Reinaud, J. 2005. 'Industrial competitiveness under the European Union Emissions Trading Scheme'. Paris: IEA Information paper.

Speck, S. 2006. 'Overview of environmental tax reforms in EU member states'. COMETR DL 1.3: Reviewed, Revised, and Condensed Research Report to provide input for DL 7.2: Part B.

Szabo, L., Hidalgo, I., Ciscar, J. C., and Soria, A. 2006. 'CO_2 emission trading within the European Union and Annex B countries: the cement industry case'. *Energy Policy*, 34: 72–87.

Part IV

Implications for Future Climate Policy

9

Carbon Taxes and Emissions Trading: Issues and Interactions

Paul Ekins[1]

9.1 Introduction

There are two principal economic instruments that have been both considered and implemented in an attempt to reduce carbon emissions: carbon (or energy) taxes, which, combined with reductions in other taxes in an environmental tax reform (ETR), are the main subject of this book; and carbon emissions trading. The European Union (EU) has set up the largest carbon emissions trading scheme in the world, covering about 50 per cent of EU carbon emissions. This chapter compares carbon taxes and emissions trading and considers interactions and potential conflicts and synergies between them.

The EU emissions trading scheme (EU ETS) was set up by European Directive (EC, 2003a) and is envisaged to run in three phases: Phase 1, now completed, from 2005–7, the current Phase 2, from 2008–11, and Phase 3 from 2012–20. The scheme is complex, and details of it may be found on the European Commission's website[2] and will not be rehearsed in any detail here.

This chapter first discusses characteristics of emissions trading schemes in general, before some theoretical discussion of the similarities and differences between carbon taxes and emissions trading. It then explores, in the European context, the possibilities and implications of introducing taxes and emissions trading alongside each other, in order to arrive at a

---

[1] Paul Ekins, Professor of Energy and Environment Policy, UCL Energy Institute, University College, London, United Kingdom.
[2] See <http://ec.europa.eu/environment/climat/emission/index_en.htm>.

view as to whether, with the EU ETS now in place, ETR might still have a role in European climate and energy policy, at an EU or member state level or both.

9.2 Emissions trading

Emissions trading involves the issuing, normally by government, of emissions permits, or allowances, to cover the desired quantity of emissions, and their transfer, by sale or otherwise, to emitters. Emitters may trade these permits among themselves, subject only to the requirement that they surrender to the authorities at the end of the relevant period, normally each year, a quantity of permits that is equal to their emissions over the period. Normally, the number of permits is lower than would have been emitted in the absence of the scheme (if this is not the case, there is no point in the scheme being introduced), so that some emitters will need to undertake abatement. The scheme is intended to ensure that abatement will be undertaken by those who can do so at the lowest cost, who will then need fewer permits to buy from the authorities (if the permits are being sold) or may have permits to sell to higher-cost emitters if the permits have been distributed on some other basis. The price of the permits will reflect both the scarcity of permits (the tightness of the emissions 'cap' that has been imposed) and the costs of abatement. For reasons that should be clear from this description, such a scheme is sometimes called a 'cap & trade' emissions trading scheme.

Figure 9.1 shows the price of carbon under the first phase of the EU ETS, and the forward prices into the second phase (CBI, 2007). The sharp fall in price in April 2006 was due to the announcement that emissions under the scheme were lower than had been expected, resulting in a glut of allowances as Phase 1 came to a close at the end of 2007, with no possibility of banking the allowances into Phase 2, so that the price fell to zero. Trading in the first half of 2007 before the start of Phase 2 saw the price of Phase 2 allowances vary between about €12 and €25/tonne CO_2. In 2008, the onset of the economic downturn saw the price fall from around €18 to €14/tCO_2. This volatility of carbon prices in a trading scheme, especially before the emissions market reaches maturity, is one of the arguments for combining trading with a carbon tax, as will be seen.

It is important to recognize that the level of emissions that arises in a trading scheme depends only on the cap that has been set. Extra policies to reduce emissions (such as, for example in the European context,

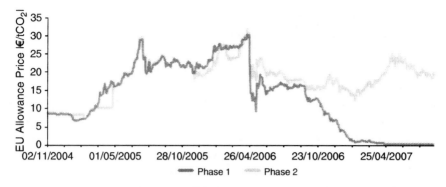

Figure 9.1. The price of carbon under EU ETS

Source: Climate Change Capital.

policies to increase the quantity of renewable electricity or to increase the energy efficiency of electricity use) will not reduce emissions below the cap but, to the extent that they are successful, will reduce the price of allowances. It is therefore very important that the impact of such policies on emissions is taken into account *before* the cap is set, so that the quantity of emissions in the cap is adjusted downwards to take these policies into account, if it is desired to maintain a robust carbon price to stimulate further low-carbon investment or induce demand reduction.

9.3 Competitiveness implications of emissions trading

It is to be expected that the limiting of carbon emissions through the imposition of a cap, and the creation of a carbon price to reflect the scarcity of the associated emission allowances, will have an effect on the competitiveness of the firms included in the EU ETS. However, as with the imposition of carbon taxes, this effect is by no means straightforward and needs careful analysis.

Most obviously, many of the same considerations in respect of carbon taxes and competitiveness are also relevant to emissions trading. Thus, the competitiveness effects may be expected to differ according to the carbon intensity of the sector, the trade intensity of the sector, the means of recycling any revenues from the sale of emission permits, the availability and managerial awareness of cost-effective technological means of carbon abatement, and managerial willingness and competence to take

243

advantage of them. These issues are extensively covered elsewhere in this book, and will not be further mentioned here.

The trading-specific issue that may be expected to have most implications for competitiveness is how the emissions permits are allocated. Broadly, there are two options. Either they may be given away for free, perhaps according to historical emissions in some particular year (called 'grandfathering'), or according to some reference emission intensity for the sector (called 'benchmarking'); or they may be sold by the government to emitters, normally by auctioning. In the EU ETS, in Phase 1, all the permits were grandfathered. In Phase 2, the great majority were grandfathered, but up to 10 per cent could be auctioned by governments. In Phase 3, it is envisaged that nearly all the permits for the power sector will be auctioned from 2013, and an increasing proportion over time for other sectors, arriving at full auctioning for the EU ETS by 2027. Competitiveness considerations played a major role in the negotiations for increasing the proportion of auctioned permits in the EU ETS, and in the shape of the final agreement that was reached.[3]

It is often not recognized that the way emission allowances are allocated in an ETS does not affect the carbon price that arises from it. Thus, it makes no difference to the carbon price whether the allowances are given away by the government or sold at auction or in some other way. The carbon price depends solely on the cap (the absolute quantity of allowances) and the costs of abatement. What the method of allocation does affect, of course, is whether it is firms (in the case of free distribution) or governments (in the case of allowance sales) who benefit from the revenues that derive from the possession of carbon allowances. When allowances are auctioned, governments benefit directly from the revenues accruing from the auctions, and this represents a direct cost to firms which may affect their competitiveness.

When emission permits are given away, the competitiveness effects on firms of a trading scheme like the EU ETS may arise via two routes. Most obviously, direct costs may arise as industries need to acquire certificates for additional production activities. Indirect effects will arise from the differential ability of firms to pass on the market price of carbon to their consumers. Even when allowances are given away for free, firms will still try to do this because, despite their free distribution, the allowances represent an opportunity cost to firms when they are set against emissions

---

[3] For further detail of this agreement, see EP (2008).

rather than sold, so normal marginal cost pricing would seek to recover the allowance cost in the price paid by consumers. To the extent that firms are able in their markets to pass on the price of the allowances to their consumers—and this will depend on whether they are predominantly price-setters or price-takers in those markets (see Chapter 4)—they can actually increase their profitability in a trading scheme with free distribution of permits. When the consumers on the receiving end of these price increases are also firms, then this can represent another competitiveness impact from the EU ETS over and above the issue of allowance allocation. The sector for which these issues have most often been raised is the power sector.

Numerous studies have investigated the pass-through of permit prices to electricity prices. The most pessimistic studies assume a 100 per cent pass-through rate; for example, McKinsey and ECOFYS (2006) estimates a figure of €10/MWh for a €20 allowance price. The Carbon Trust (2004: 11) has estimated that the electricity sector is able to pass through 90 per cent of the allowance price to its consumers by increasing the price of electricity, substantially increasing its profits (even more so if this increase applies whether or not the electricity is generated from carbon-based fuels).

However, several studies show that the pass-through rate will only be 100 per cent during the time when power demand exceeds the base load, and where it is coal or lignite plants that set the marginal price (Sijm et al., 2006). In the seasons and periods where hydropower or nuclear power sets the marginal price, it is not likely that power operators will be able to factor in the full value of the certificates. One study, for Germany and The Netherlands, comes to a pass-through rate of 40–60 per cent. The IEA furthermore points out that large parts of the European electricity market are not yet fully liberalized and that price regulations will restrict pass-through (Reinaud, 2007). Nevertheless, the IEA points to the Nordic electricity market (Nordpool) as one region where the electricity trade has been successfully liberalized and where pass-through of ETS costs should be expected. Due to the significance of hydro- and nuclear power, one Finnish study concludes that the average pass-through rate on the Nordpool exchange should be in the range of 40 per cent, for example, €4/MWh for a €20 allowance price (VATT, 2006). This would mean that the range indicated by the various studies and market analysts runs from €4–10/MWh for the power sector with a €20 allowance price. This pass-through cost can be compared with CO_2 tax rates on electricity in the range of €6–12/MWh for smaller business users in The Netherlands, UK,

Germany, and Denmark, and €0.5/MWh for large users (energy-intensive industries).

In conclusion, in respect of the power sector, the ETS system will from 2008 most likely effectively increase the costs per kWh for businesses in EU27 to a level comparable to the ETR-induced electricity price rises for smaller business users in the ETR countries, which will go some way to offsetting the fiscal exemptions obtained by energy-intensive industries under ETR. However, as there is no revenue available for recycling under the grandfathered ETS scheme, no simple way exists in which to compensate energy-intensive industries for the additional burden, for example by lowering employers' social security contributions. Hence it is possible that there are more substantial inroads into energy-intensive industries' gross operating surplus from ETS than from the pre-existing carbon-energy taxes introduced under ETR. However, as fuel uses other than electricity are not affected by pass-through, the overall impact is still likely to be less than that of carbon-energy taxes levied on all fuels. Estimates suggest that aggregate profits in the electricity sector of just Belgium, France, Germany. and The Netherlands from passing through the price of grandfathered permits could total €5–14 billion, depending on the assumed model of competition and elasticity of demand (Sijm *et al.*, 2006: 62). Auctioning the allowances would transfer these profits to the government, but would hardly affect the price of electricity.

Other sectors, in particular those that operate in competitive global markets, are much less able to pass on the carbon price to their consumers, and the competitiveness of these sectors is more likely to be affected when permits are auctioned rather than given away. EC 2008 (p. 111) noted that there is no definitive indicator of the ability to pass through costs of emission permits, but one relevant consideration is obviously 'openness to trade'. Figure 9.2 shows the openness to EU trade of various industrial sectors, many of which are participants in the EU ETS, showing that non-ferrous metals and chemicals are among the most open sectors, and non-metallic minerals among the least open. While ferrous metals appear among the less open sectors, it would be wrong to conclude that therefore this sector is relatively less vulnerable to competitive pressures, because of other factors, as discussed in Chapter 4.

However, as with carbon taxation and ETR, the competitiveness effects of an ETS with auctioned permits will depend to a great extent on how the resulting revenues are recycled through the economy. It is envisaged that in 2020 allowances for 1,720 million tonnes CO_2 will be issued under the EU ETS (EP, 2008). If all these were to be auctioned

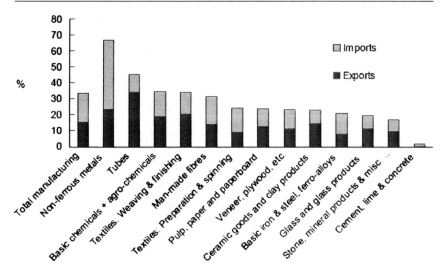

Figure 9.2. Openness to extra-EU trade, 2004–2005

Source: Eurostat Prodcom and UN Comtrade databases, cited in EC (2008: 112).

at a price of €20/tonne CO_2 (in practice, a relatively small proportion would still be grandfathered at this date), this would result in revenues of around €34 billion, of which it is envisaged that around 90 per cent would be retained at member state level, of which around half should be spent on climate mitigation and adaptation measures (although this is not mandatory). Of course, not all the revenues would be additional to current government revenues, because firms would be able to offset the purchase of permits against other tax liabilities. This suggests that member states' revenues in 2020 could be augmented by at least €20 billion, which is a sum comparable to the revenues of the ETRs examined in this book. The competitiveness effects of the EU ETS will be greatly influenced by how these revenues are recycled through the member states' economies, concerning which the experiences of the ETRs discussed in this book are very relevant.

In its Impact Assessment for its proposals for Phase 3 of the EU ETS (EC 2008), the European Commission used the E3ME model (the same model used in the COMETR project, with results reported in Chapters 7 and 8) to estimate the impacts of full auctioning in the EU ETS by 2020. It found that full auctioning had a negative impact on GDP of 0.1 per cent (EC 2008: 114) by that date, compared to the generally positive effects on GDP of the European ETRs presented in Chapter 8, although all the effects

247

Table 9.1. Sectoral impacts on the output of different industrial sectors from full auctioning of permits in the EU ETS by 2020

SECTOR	Percentage Difference From Baseline
Agriculture and mining	0.2
Basic metals	0.0
Non-metallic minerals	0.0
Wood and paper	−0.1
Chemicals	0.4
Rubber and plastics	0.2
Air transport	0.9
Electricity	−0.1
Non-ETS manufacturing	0.1
Construction	−0.1
Services	−0.1

Note: Figures show percentage difference from baseline at EU25 level in 2020 from auctioning all EU ETS allowances in 2020.
Source: E3ME model, cited in EC (2008: 114–15).

are rather small, and care must be taken with such comparisons because of differences in the baselines and scenario specifications. The Impact Assessment also modelled the impacts of full auctioning on different economic sectors, and the results of this are shown in Table 9.1, from which it can be seen that the impacts in different sectors are both positive and negative, but again they are rather small, in all cases being below 1 per cent of sectoral output. Again, these results provide additional insights into the sectoral results from the COMETR project reported in Chapters 4, 5, 6, and 7.

9.4 Carbon taxes and emissions trading

In any market, attempts to fix both the price and the quantity will fail. In respect of carbon, a carbon tax will cause emitters to reduce their emissions through abatement to the extent that their marginal abatement cost is below the tax rate, so the quantity of emissions then depends on the opportunities for and costs of abatement. Emissions trading, in contrast, fixes the quantity of emissions, and abatement takes place to reduce emissions to that quantity, such that the price of carbon becomes equal to the marginal cost of abatement at that quantity.

As discussed in Ekins and Barker (2001), it has been shown, under a precise set of restrictive assumptions, that there is broad equivalence between an emissions trading scheme, where emission permits are auctioned by the government, and levying a carbon tax at the auction price (Pezzey, 1992; Farrow, 1995). Under such conditions, therefore, the main difference between taxation and trading concerns price/quantity adjustment. With a carbon tax, it is the tax on, and hence the price of, carbon that is fixed, and the quantity of carbon emitted as CO_2 that adjusts. With emissions trading, it is the quantity of carbon emitted as CO_2 that is fixed, and the price of the emission permits that adjusts.

With regard to instrument choice, Weitzman (1974) has shown that (1) it is preferable to fix the price through a tax when there is uncertainty over the abatement cost function, and a possibility that it is very sensitive upwards to greater than optimal carbon emissions reduction, and (2) it is preferable to fix the quantity through a cap when there is uncertainty about the damage function, and a possibility that it may be very sensitive upwards to greater than optimal emissions (for example, through the existence of climatic 'tipping points'). Using this insight, Pizer (1999) has argued that it would be preferable to control carbon emissions using a price, rather than a quantity, instrument, in contrast to the provisions of the Kyoto Protocol, which are for quantity control. Pizer suggests that the problems in negotiating the details of the Protocol derive from the potentially high costs which carbon limits may entail. These costs could themselves be limited by specifying a 'trigger price' for extra emission permits, which would effectively set a maximum cost of abatement (though obviously emissions could then increase). The proposal well illustrates the relation and interaction between prices and quantity limits.

Hepburn (2006: 238) agrees that 'unless we are certain that we are on the brink of a tipping point, a carbon tax appears superior to tradable quantities', but he also recognizes that the establishment of the EU ETS and the provisions in the Kyoto Protocol for mechanisms of trading rather than taxation mean that it is unlikely that carbon taxation will now replace trading as the international system of carbon emissions reduction. This leads to the question of whether carbon taxation might have a complementary role to trading and be introduced alongside it, rather than seeking to replace it.

9.5 The interactions between taxes and trading

As seen in Figure 9.1, one of the characteristics of the EU ETS to date has been considerable volatility in the carbon price it produces.

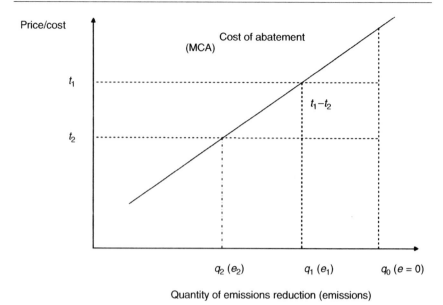

Figure 9.3. The interaction between a carbon tax and emissions trading system

Such volatility introduces considerable uncertainty and risks into the prospects for low-carbon investments, which can only serve to discourage such investment. Since one of the purposes of the carbon price is to act as a stimulus to low-carbon investment, the volatility is clearly undesirable.

One way of addressing this volatility, as noted by Hepburn (2006: 230) would be for the government to place a floor on the price, by guaranteeing to buy any permits offered for sale at this price, or to place a ceiling on the price either by offering to sell an unlimited number of permits at the ceiling price, or, equivalently, by allowing trading participants to pay a penalty, or 'buy-out' price, for any emissions for which they had no allowances.[4]

Another way of setting a floor on the price of carbon emission permits would be to set a carbon tax at the level of the desired floor price. Figure 9.3 sets out how this might work. The horizontal axis shows the quantity of emissions reduction (quantity of emissions) and the vertical

---

[4] The UK Renewables Obligation works like this by allowing electricity suppliers to pay a 'buy-out' price to cover any shortfall in their purchase of Renewable Obligation Certificates (ROCs).

axis the price of carbon or the cost of abatement. The marginal cost of abatement (MCA) rises with the level of emissions reduction. At q_0 emissions have fallen to zero.

Imagine a trading scheme that sets the quantity of emissions reduction (below some baseline) at q_1, corresponding to emissions of e_1. Then this yields a price of carbon of t_1, corresponding to the MCA needed to reduce emissions by q_1. Because MCA is uncertain in a new market, this price might be volatile, as shown in Figure 9.1. Then imagine that a carbon tax is introduced at level t_2. By itself this would lead to a quantity of emissions reduction of q_2 ($<q_1$). However, the total quantity of emissions reduction remains at q_1, because that is set by the cap. What adjusts is the price of permits. Without the carbon tax, the permit price is t_1. With a carbon tax of t_2, the permit price falls to $t_1 - t_2$, so that the overall cost of carbon remains the same at t_1. The difference is that the tax effectively sets a floor on the carbon price of t_2. If the carbon price in the permit market would tend to fall below this, then the permit price would fall to zero, but all emitters of carbon would still need to pay the carbon tax of t_2. An advantage of the tax in this context would be that it would give assurance to low-carbon investors of a minimum carbon price. This could be gradually moved upward over time in a predictable way to give further assurance about the returns from these investments. A further advantage of the tax is that it would guarantee a floor not only on the carbon price, but on the revenues for the government to be derived from it, which is not the case with auctioning and price volatility.

It may be noted that a carbon tax would only affect the price of permits if it were introduced at an EU level (or in a significant enough number of member states to have an impact on the EU carbon market). A national carbon or energy tax, such as the UK climate change levy or some of the other taxes involved in the ETRs of some EU countries discussed in this book, which fell on carbon emissions or energy use that were already covered by the EU ETS, would neither reduce the price of permits, nor reduce the quantity of emissions (which is determined by the EU ETS cap). Such a tax would simply represent an extra cost for the industries affected by it, as well as being a revenue source for the government.

9.6 Carbon taxes and emissions trading in the EU

In the EU, taxation is considered to be a competence of member states, rather than of the European Commission, so that unanimity on tax proposals put forward by the Commission is required. In the 1990s, there were strenuous efforts by the Commission to persuade member states to agree to an EU-wide carbon-energy tax, but the proposal failed to achieve unanimous acceptance and it was never accepted. Following the agreement of the Kyoto Protocol in 1997, the Commission turned its attention to emissions trading, to match the flexible mechanisms that were envisaged in the Protocol, and the EU ETS was formulated. Its acceptance in 2003 represented a remarkable policy achievement and it is widely regarded as a path-breaking, essential foundation for a global emissions trading scheme for climate change mitigation in the future. However, it has not been without its problems, as noted above, including a volatile price and, because the great majority of allowances were given away, the generation of 'windfall profits' for those sectors which, like electricity, were able to pass on all or most of the price of the allowances to their customers. One result has been that, as noted above, a far greater proportion of the allowances for Phase 3 of EU ETS, due to begin in 2013, will be auctioned, moving EU ETS closer to being a tax.

At the same time as working on EU ETS, and following the agreement in 1992 of common rules for taxation of mineral oils, the European Commission developed the Energy Tax Directive, which was finally unanimously agreed in 2003 (EC, 2003b), and which set minimum rates of excise duty for motor fuels, heating fuels, and electricity, albeit with very low rates and with a very large number of reduced rates and exemptions for different member states. Even so, the joint existence at the EU level of Directives on both taxation and emissions trading has raised concern about double regulation.

The ETS system divides stationary emitters into two sectors: ETS and non-ETS. The argument in support of the double-regulation concern is that as emissions from the ETS sector are fully regulated from the trade with certificates, there is no further need for a regulatory tax. The ETS sets a cap for emissions from the ETS sector; if emissions exceed this cap, additional allowances must be acquired on the market, possibly with the use of other flexible instruments as well. However, there are also issues as to whether an ETS grandfathered allowance qualifies as a fully fledged scheme of tradable quotas; and the fact that the Energy Taxation Directive has a broader scope than carbon taxation, being also concerned with security of energy supply and harmonization of tax rates. As the ETS system has created a market where prices have shown to be very volatile, and as pass-through rates are very dependent on the specificities of electricity and other markets, the ETS system will not be likely to create a level playing field, as was the intention of the minimum tax rates, and there are concerns about the implications for the polluter-pays principle.

Proposals for a revision of the Energy Tax Directive are expected to be put forward in early 2009, and a Green Paper outlining the European Commission's initial thoughts was issued in 2007 (EC, 2007). One of these thoughts raises the possibility of splitting the minimum excise duty into energy and carbon components, which would effectively introduce a minimum carbon tax across the EU, albeit at very much lower levels than were proposed in the 1990s. This could be one way of addressing the double-regulation concern, because it would allow EU ETS sectors to be exempted from the carbon component of the tax, in recognition of the fact that this component was already covered by the EU ETS, but not from the energy component, in recognition of the Energy Tax Directive's other objectives. However, it is not yet clear whether this proposal to split the minimum energy tax rates into these two components would win the unanimous agreement from member states which it would need to become operational.

Although it currently seems most unlikely that there would be sufficient agreement in all member states for this revision to result in the re-submission of a proposal for an EU-wide carbon tax at a significant level, there are good reasons for such a move. First, it would set a floor on both the carbon price in the EU ETS and the revenues from it for government, as discussed above. This would complement the increased auctioning that is envisaged for Phase 3 of EU ETS. Guidance on an appropriate level for the tax would be given by the carbon price in the rather limited auctioning of allowances in EU ETS that is taking place in Phase 2. Secondly, it would cover all the other carbon emissions that are currently not covered by the EU ETS, and ensure that they too were subject to a minimum price across the EU beyond the very low current prices of the Energy Tax Directive. Thirdly, the tax could be increased gradually over time to help achieve the ambitious carbon reduction goals across the EU in a relatively harmonized way (individual member states could of course introduce supplementary measures of their own, which might or might not include carbon taxes or trading schemes). With the revenues from the EU carbon tax being retained at the member state level, this would give an opportunity for member states to enact ETRs by reducing other taxes, in the knowledge that, because the carbon tax was being implemented EU-wide, many of the potential competitiveness effects, fears of which have produced the very complex tax provisions for energy-intensive industrial sectors described in Chapter 2, would not arise.

These are powerful arguments for the introduction of a significant EU carbon tax alongside the EU ETS. One way in which the force of these arguments could perhaps be better disseminated is through the establishment of an EU forum on market-based instruments, as tentatively proposed in the European Commission's Green Paper on the revision of the Energy Tax Directive (EC 2007: 5). However, as past experience shows, the obstacle posed by the unanimity requirement for new EU tax proposals means that the challenge of winning approval for significant new EU tax initiatives should not be underestimated.

References

Carbon Trust. 2004. 'The European Trading Scheme: Implications for Industrial Competitiveness'. July. Carbon Trust, London.

CBI (Confederation of British Industry). 2007. 'Climate Change: Everyone's Business'. November. CBI, London.

EC (European Commission). 2003a. Directive 2003/87/EC of the European Parliament and of the Council of 13 October 2003 establishing a scheme for greenhouse gas emission allowance trading within the Community and amending Council Directive 96/61/EC. OJ L275, 25 Oct., pp. 32–46.

——2003b. Council Directive 2003/96/EC of 27 October 2003 restructuring the Community framework for the taxation of energy products and electricity. OJ L283, 31 Oct., pp. 51–70.

——2007. Green Paper on market-based instruments for environment and energy related policy purposes. COM(2007) 140 final. Brussels: Commission of the European Communities.

——2008. Accompanying document to the Proposal for a Directive of the European Parliament and of the Council amending Directive 2003/87/EC so as to improve and extend the EU greenhouse gas emission allowance trading system: impact assessment. 23 Jan. Brussels: Commission of the European Communities. <http://ec.europa.eu/environment/climat/emission/ets_post2012_en.htm>.

Ekins, P. and Barker, T. 2001. 'Carbon taxes and carbon emissions trading'. *Journal of Economic Surveys*, 15/3: 325–76.

EP (European Parliament). 2008. European Parliament legislative resolution of 17 December 2008 on the proposal for a directive of the European Parliament and of the Council amending Directive 2003/87/EC so as to improve and extend the greenhouse gas emission allowance trading system of the Community. <http://www.europarl.europa.eu/sides/getDoc.do?pubRef=-//EP//TEXT+TA+P6-TA-2008-0610+0+DOC+XML+V0//EN&language=EN>, or <http://europa.eu/rapid/pressReleasesAction.do?reference=IP/08/1998>.

Farrow, S. 1995. 'The dual political economy of taxes and tradable permits'. *Economic Letters*, 49: 217–20.

Hepburn, C. 2006. 'Regulation by prices, quantities or both: a review of instrument choice'. *Oxford Review of Economic Policy*, 22/2: 226–47.

McKinsey and ECOFYS. 2006. *EU ETS Review: International Competitiveness*. Brussels: European Commission DG ENV, 51 pp.

Pezzey, J. 1992. 'The symmetry between controlling pollution by price and controlling it by quantity'. *Canadian Journal of Economics*, 25/4: 983–91.

Pizer, W. 1999. 'Choosing price or quantity controls for greenhouse gases'. Climate Issues Brief No.17, Washington DC: Resources for the Future.

Reinaud, J. 2007. 'CO_2 allowance and electricity price interaction'. IEA Information Paper. Paris: International Energy Agency.

Sijm, J., Neuhoff, K., and Chen, Y. 2006. 'CO_2 cost pass-through and windfall profits in the power sector'. *Climate Policy*, 6: 49–72.

VATT (Government Institute for Economic Research). 2006. *Impacts of the European Emission Trade System on Finnish Wholesale Electricity Prices*. VATT Discussion Papers, 405. Helsinki.

Weitzmann, M. 1974. 'Prices vs. quantities'. *Review of Economic Studies*, 41: 477–91.

10

Conclusions: Europe's Lessons from Carbon-Energy Taxation

Mikael Skou Andersen[1] *and Paul Ekins*[2]

10.1 Introduction

In this chapter we provide an overview and interpretation of findings presented in this volume, while placing them in the context of the wider climate policy debate. We return to our point of departure: the idea that properly designed carbon-energy taxes, in addition to lowering emissions, may allow for a second dividend by way of improved employment and economic performance. First, we summarize the serious challenge facing the world today of the need to stabilize atmospheric greenhouse gas emissions and we examine some of the cost estimates available for mitigation of global warming. Secondly, we revisit the competitiveness issue and consider the findings presented in this book within the framework of the Porter hypothesis and Leibenstein's concept of X-efficiency, both of which have been quoted in support of more vigorous energy and climate policy. In so doing, we pay particular attention to energy-intensive industries, such as basic metals, cement, aluminium, and chemicals, and discuss our findings in view of other recent studies that have addressed the potential for improved energy productivity and fuel shifts. Carbon leakage, which refers to the displacement of emissions to non-carbon tax countries and regions, is a prominent concern in relation to these industrial sectors and we examine the leakage rates identified here against the broader patterns

[1] Mikael Skou Andersen, Professor, National Environmental Research Institute, Aarhus University, Denmark.
[2] Paul Ekins, Professor of Energy and Environment Policy, UCL Energy Institute, University College, London, United Kingdom.

of development in international trade and development, with particular attention directed towards developments in China and other emerging industrialized countries. While carbon leakage remains a risk as long as the global political framework for controlling emission of greenhouse gases is incomplete, the obvious challenge is to identify a formula that enables control, while at the same time allowing for transformation of global energy systems and continued economic growth, in particular in developing countries, where poverty alleviation should also be a priority. These considerations lead to a discussion of border-tax arrangements in relation to carbon-energy taxation in general. We conclude that not only do the well-established approaches to carbon-energy taxation as seen in Europe—in the form of revenue-neutral tax-shift reforms—deserve attention, but so also do solutions of a more tailor-made design for the emerging economies (in particular in Asia). Competitiveness concerns, of central importance to policy-makers, will need to be realigned with those of fairness and equity that figure prominently in emerging economies and which to date are not party to the control mechanisms under the UN Climate Convention. Against this background, the analysis leads to proposed use of an escalator to increase carbon-energy taxes according to differentiated responsibilities.

10.2 Avoiding dangerous climate change

10.2.1 *The greenhouse gas stabilization challenge*

The best estimates for continued, 'business-as-usual' development of the global economy, with a world population that reaches 9 billion, project that global emissions of greenhouse gases are likely to have doubled by 2030. As CO_2 is a gas that accumulates in the atmosphere, this would lead to a CO_2-equivalent concentration in the region of 500 ppm (parts per million), close to double the 280 ppm level that prevailed before humans began to use fossil fuels. On the basis of comprehensive climate modelling, as published in the scientific literature, the Intergovernmental Panel on Climate Change (IPCC) has warned that a doubling of the greenhouse gas concentration would very likely lead, by 2050, to a temperature increase of about 2.3 degrees C. By the end of the present century, temperature changes are more uncertain, but could well involve increases of between 3 and 4.5 degrees C. Evidence from the March 2009 Copenhagen climate conference suggests that, even since the IPCC's 2001 report, there are substantially increased reasons for concern about the impacts of projected climate change on human health and welfare in this century (see Richardson et al. 2009, Figure 8, p. 16).

Official policy in the European Union is to stabilize the global temperature increase at 2 degrees. The rationale behind this policy stems partly from a desire to limit the impacts of climate change, but also from what is judged to be realistic. The 2-degree C increase is an increase over global temperature in pre-industrial times—however, since this time, temperature is believed already to have increased by around 0.7 degrees C. A further commitment of 0.6 degrees C is moreover embodied in the concentration of greenhouse gases currently held in the atmosphere. As such, the room for manoeuvre within the 2-degree C target has already narrowed significantly, by 1.3 degrees C, and unless emission increases immediately cease, will continue to do so.

A global temperature of 2 degrees C over that experienced today is not without precedent. In the Eem period, around 125,000 years ago, climatic records indicate that temperatures were approximately 2 degrees C higher than at present—due to variations in the amount of solar energy reaching the Earth (variations caused by changes in the solar cycle, e.g. the Milankovich effect). During the Eem period, glacial cover was less than at present and sea water levels approximately 4–5 m higher. Although uncertainties remain, the fundamental patterns of these historical circumstances can be captured by the climate models. With a global temperature rise of more than 2–3 degrees, the risk of more uncontrolled and abrupt climate change is believed to be significant. At such levels of climatic change, well-known processes in the soil could lead to the release of large amounts of methane, a greenhouse gas with a much higher impact and radiative potency than CO_2. The choice of a 2-degree target is therefore not without justification, although accounting for the risks with any great precision is evidently problematic.

As inventories of greenhouse gas emissions have been extended across the globe, the basis for projecting the implications of the various reduction and stabilization trajectories has improved. Figure 10.1, based on modelling from the Climate Research Centre in Potsdam (den Elzen and Meinshausen, 2006; Meinshausen, 2006; Meinshausen et al., 2006), provides an overview of the relationships believed to exist between the 2-degree greenhouse gas stabilization target and the chances of attaining this target under different emission scenarios. This figure indicates that there is a pressing need for significant emission reductions within a short time frame. To retain a 50/50 chance of staying within the 2-degree warming, the world would need to stabilize at a concentration as low

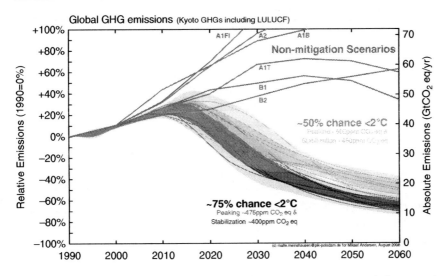

Figure 10.1. Scenarios for future global greenhouse gas emissions and chances of not exceeding global warming of 2 degrees Celsius

Source: Meinshausen *et al.*, (2006)

as 450 ppm CO_2-equivalents. Stern (2007: 193) notes that the global atmospheric concentration of greenhouse gases (not just CO_2) was then 430 ppm, and rising at 2.5 ppm per year, at which rate, the 450 ppm concentration will be reached by about 2015. Some overshooting seems inevitable. A stabilization requirement of 450 ppm is therefore a very demanding target, and well below the above-mentioned doubling at 550 ppm. Global emissions would have to begin to decline rather than grow within the next ten years, and over the next decades a reduction of 30–40 per cent (relative to 1990 levels) would be required. A greater chance (75 per cent) of accomplishing the 2-degree target requires that earlier action is taken and also that more significant reductions take place in the decades to follow.

According to climate scientists, it is these implications of the relentless laws of physics which set the frame for policy-making aimed at containing climate change. While modelling future temperature increases is difficult, and sensitive to key assumptions, it is worth noting that the paleoclimatic record of past temperatures and carbon concentrations provides fairly robust support both for the basic problem of understanding the climate system and for calibration of the climate models.

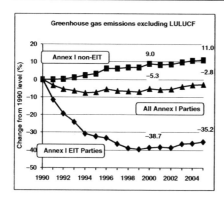 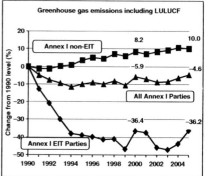

Figure 10.2. Greenhouse gas emissions from Kyoto Protocol Annex 1 countries, split into economies in transition (EIT) and non-EIT.

Source: Kononov, (2007).

10.2.2 The costs of mitigation

Controversy surrounding the costs of the Kyoto Protocol to the UN Framework Convention on Climate Change was to some extent amplified by an unfavourable relationship between, on the one hand, the protocol's limited reduction impact on global warming and, on the other hand, costs that were often estimated without reference to opportunities for technological change. The Kyoto Protocol aims to stabilize *emissions* of greenhouse gases (GHGs) not *concentrations*, and it covers only developed countries, there being no binding commitments for developing countries. The limited scope of the Kyoto Protocol, and the circumstance that the biggest emitter, the USA, has not yet ratified, implies that the protocol should be regarded only as an intermediate measure towards more comprehensive emission control among all emitters. Nevertheless, the Kyoto Protocol has been strategically important in establishing an institutional architecture for international cooperation on mitigation of greenhouse gas emissions, and wider international cooperation would be difficult to conceive had the protocol not been in place.

On aggregate, the overall reduction in emissions aimed for in the Kyoto Protocol seems likely to be achieved, even in the absence of a commitment from the USA (Figure 10.2; cf. UNFCC, 2007; Kononov, 2007). Emissions reductions in former planned economies in relation to a 1990 baseline have been considerable and have more than offset an increase in emissions in the USA. The European Union has managed to stabilize emissions; although this is also due in part to 'hot air' (differences between present emissions and those in 1990, before the economic transition in Central and Eastern Europe) produced by new EU member states with economies in transition.

Barker and Ekins (2004: 68) discuss how cost estimates relate to the modelling approaches and assumptions employed, concluding the high-cost estimates in the literature are either misleading or they demonstrate

1. the costs of making policy mistakes (e.g. by too hasty, unexpected action, or sub-optimal use of the revenues)
2. the costs of policies that do not include the use of the Kyoto flexible mechanisms and/or
3. how a selection of worst-case assumptions and parameters can accumulate to give high costs.

Taking all the literature into account the macroeconomic costs of greenhouse gas mitigation of the kind envisaged by the Kyoto commitments is likely to be insignificant in the US, provided that the policies are expected, long-term and well-designed.

Looking forwards to the very large GHG emission reductions that will be required to achieve atmospheric stabilization at 450–550 ppm, there is a wide range of estimates of the associated macroeconomic costs.

Figure 10.3 shows a scatter plot of the macroeconomic cost estimates from a large number of different models and model runs for varying reductions in CO_2 emissions by 2050, which was part of the evidence base that caused the Stern Report to conclude that stabilization of atmospheric concentrations of GHGs at 500–50 ppm could be kept at or below 1 per cent of GDP by mid-century (Stern 2007: 276). In a similar comparison of a range of estimates, the IPCC's Working Group III found that the median cost estimate of stabilizing GHGs at 535–90 ppm CO_2 was 1.3 per cent GDP (IPCC 2007a: 27). In an economy growing at 2 per cent per year, such a reduction implies that stabilization would simply mean the deferral by one year of any given year's level of consumption.

Macroeconomic costs expressed in percentages of GDP are very different in terms of their political resonance to the possible effects on the economic competitiveness of different sectors, which have been the principal focus of this book. The next section discusses the role of

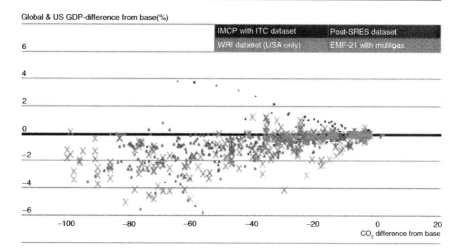

Figure 10.3. Projections of macroeconomic costs (as a share of GDP) associated with different CO_2-reduction targets—scatter-plot of modelling results

Note: (1) Each point refers to one year's observation from a particular model for changes from the reference case for CO_2 and the associated change in GDP from four sources for years over the period 2000–2050.

Source: Stern (2007:270), citing Barker *et al.*, (2006).

carbon-energy taxation in reducing GHGs, before summarizing the evidence on competitiveness from the COMETR project in the section following.

10.3 Greenhouse gas reductions: the role of carbon-energy taxation and emissions trading

10.3.1 The effectiveness of carbon-energy taxation in emissions reduction

Modelling with E3ME (Barker *et al.*, Chapters 7 and 8, this volume) indicates that carbon-energy taxes, introduced under environmental tax reform in Europe, caused reductions in greenhouse gases in the range of 4–6 per cent for the Nordic countries and Germany in the decade from the mid-1990s up to 2004, mainly as a result of reductions in fuel demand. These are reductions in relation to the reference scenario of business-as-usual without any tax reform, not absolute reductions. For the UK and The Netherlands, reductions were lower, approximately 2 per cent, as a result of later and less ambitious programmes of tax shifts. Impacts of the tax shifts will stabilize in the years up to 2012; for Sweden, however, a reduction of up to 7 per cent is projected as a result of continued tax shifting. For the countries which introduced environmental tax reforms, CO_2 emissions are projected to have been reduced overall by approximately 60 million tonnes of CO_2 by 2012 in relation to a business-as-usual scenario without these taxes. This reduction will contribute towards meeting the EU15 Kyoto target of an 8 per cent reduction or approximately 315 million tonne CO_2 less than in 1990.

More detailed analysis indicates that substantial greenhouse gas reductions underlie the aggregate figures for the countries as a whole. Enevoldsen (2005) found that Danish industry improved energy productivity by around 30 per cent in the first decade of carbon-energy taxation, the improvement being a factor of two to three times higher than for industry in comparable European countries without carbon-energy taxation.

Denmark's environmental tax reform focused particularly on industry, the household and transport sectors already having been subject to significant energy taxation. Analysis of responses to carbon-energy taxes introduced in Nordic countries has shown that elasticities for industrial energy demand are in fact relatively similar for Denmark and Sweden (Enevoldsen *et al.*, 2007), which confirms the reliability of the findings for Denmark. Because Sweden has taxed industrial energy consumption since 1974, the *relative* impact of its ETR on industry fuel demand and greenhouse gas emissions has been more moderate than in Denmark. In Sweden, a more important contribution to reducing greenhouse gases came from the district heating sector, in part as a result of fuel shifts (Andersen *et al.*, 2001).

Germany's environmental tax reform assigned a high share of the carbon-energy tax burden to the transport sector, notably by increasing the gasoline tax, and to a lesser extent to industry; whereas households were affected by a moderate end-user electricity tax. The aggregate results for greenhouse gases are explained mainly by reduced energy demand in the transport sector and in certain energy-intensive industries (see also Knigge und Görlach 2004).

A particular aspect of Denmark's carbon-energy taxation programme during the first five years was the earmarking of 20 per cent of the revenue to co-finance energy-efficiency measures and upgrade production technologies. And under a programme for savings supervised by the Danish Energy Agency, businesses are required to set up an energy management system, including an energy audit procedure, staff training in energy matters, procurement policies favouring energy efficiency, and annual progress reports. An enterprise which commits to an energy-savings target and enters into an agreement with the agency becomes eligible for a partial reimbursement of the energy tax. This practice continues today. Bjørner and Togeby (1999) found that companies participating in the programme accomplished on average 60 per cent greater energy savings than companies subject to the tax only, confirming that this feature of Denmark's programme contributes to the marked impacts on energy productivity. The UK's climate change levy (CCL) offers a comparable arrangement; although only 5 per cent of the revenue is returned for energy productivity improvements, there is also the possibility for industries to reduce the tax burden under a system of agreements and hence to recycle the revenues internally for energy efficiency. A very strong announcement effect, more than twice that which could be expected from the price effect alone, was observed in relation to the introduction of the CCL, in particular for commercial energy users (Agnolucci et al., 2004). Whether the concerted efforts between industry and their energy advisers under the CCL scheme will result in similar energy productivity improvements as in Denmark remains to be seen.

In view of the need for massive reductions in greenhouse gases these effects deserve attention as pilot schemes or experiments that provide an insight into the mechanisms of carbon-energy taxation.

10.3.2 Carbon-energy taxation and emissions trading in the EU

The EU sought to introduce an EU-wide carbon-energy tax in the early 1990s, but the proposal failed to win the necessary unanimous agreement. As discussed by Ekins (Chapter 9, this volume), the EU then turned its attention to achieving minimum energy tax rates across Europe, an effort which culminated in 2003 in the passing of the Energy Tax Directive, and, after the agreement of the Kyoto Protocol, to the development of the EU Emissions Trading Scheme (ETS), the first phase of which became operational in 2005.

In Phase 1 of the EU ETS, the emissions allowances were given to companies free of charge, which effectively neutralized concerns about competitiveness. However, with the proposals in Phase 3 to auction a large proportion of emission permits, competitiveness concerns in respect of energy-intensive industries have reasserted themselves strongly, such that the original proposal for 100 per cent auctioning by 2020 has been watered down significantly, in much the same way as, under carbon-energy taxation, energy-intensive companies were regularly given tax rebates or other special considerations, as discussed in detail by Speck and Jilkova (Chapter 2, this volume).

Some EU countries had introduced environmental tax reforms around carbon-energy taxation before the failure of the EU carbon-energy tax proposal, perhaps incorrectly anticipating that it would get through. Others introduced ETRs thereafter because of the perceived effectiveness of ETR in reducing emissions in a cost-effective way. Now, again as noted by Ekins (Chapter 9, this volume), the European Commission is considering through its reform of the Energy Tax Directive reintroducing the carbon component of energy taxation. In the single country context, the issue of the competitiveness of energy-intensive industrial sectors has played an important role in both the debate about ETR and in the way it has been implemented, and this will certainly also be the case with any proposals to reform the Energy Tax Directive, especially if a significant increase in the tax rate is envisaged. The evidence with respect to competitiveness and the single-country ETRs, which has been the focus of this book, is therefore of crucial importance in terms of how both carbon-energy taxation and emissions trading in the EU are likely to develop. Considerations with respect to competitiveness relate to both economic performance—the competitive disadvantaging of home industries and their potential bankruptcy or relocation abroad—and environmental performance, through so-called carbon leakage, whereby the emissions reduction from the policy measure at home is partially or wholly offset by emissions increases abroad. The next two sections of this Conclusion respectively review the evidence from this book on these two issues.

10.4 The economic dimension of the competitiveness debate: market shares and unit energy costs

Modelling with E3ME indicates that the tax shifts undertaken in six ETR countries, whereby carbon-energy taxes substituted for other taxes, did not have negative results on overall economic performance (Barker *et al.*, Chapter 7, this volume). In the ex-post analysis, it was found that by 2012 five of the countries will have experienced a small positive gain in their GDP. Sweden faced some difficulties in the initial period, mainly due to the high initial level of carbon-energy taxation, but these economic difficulties were overcome as tax rates were adjusted to the levels of neighbouring countries. For the UK, there is no conceivable macroeconomic impact due to a late and very limited tax shift. The overall macroeconomic impact for the other five countries is moderate, reflecting that only limited revenue, as a share of GDP, has been shifted. The positive impact on GDP is less than the economic growth usually recorded for one quarter of a year. As any tax increase will normally produce a negative impact on GDP, the results here hinge on tax shifts being strictly revenue neutral.

How far do the results provide support for the claim that environmental tax shifts are capable of producing a double dividend? An ex-post analysis of the type undertaken using E3ME differs fundamentally from the ex-ante modelling normally undertaken with macroeconomic models, in that historical data are being used for the baseline scenario and that actual economic performance is used to calibrate the model. By removing the tax burden from the baseline scenario, the counterfactual is arrived at, and the difference between the two scenarios represents the disentangled impact of the taxes. Furthermore, unlike the more parameterized computable general equilibrium models, which often only have data for a single year, a macroeconomic model such as E3ME is based on historical time series that describe actual economic performance and econometric equations for the relationships between the main variables. In addition, E3ME has high sectoral disaggregation, with more than 40 sectors of the economy, and includes 12 different fuels, allowing for a realistic and detailed modelling of demand and substitution effects in the energy sector. The econometric grounding of the model gives it a better capability in representing and forecasting performance in the short to medium run. It therefore provides information which is closer to the time horizon of many policy-makers than pure CGE models.

While E3ME is a model of the entire European economy (for the EU25), other studies under COMETR have focused mainly on industrial sectors, leaving the transport sector and households aside. Where competitiveness is the issue, it is useful to focus on the manufacturing sector, and this can support a better understanding of some of the more detailed mechanisms and factors at play at the disaggregated level. In the following, we summarize what we regard as the main findings and insights achieved in the COMETR sectoral analyses.

Pricing power in the market is not always sufficiently appreciated when modelling impacts of carbon-energy taxes. Where carbon-energy taxation is introduced, manufacturing industries may avoid behavioural change or inroads into their profits if they are able to pass on, at least in part and for some markets, their cost increase. The extent to which cost increases can be passed on will reflect the sector's competitive position and strength—its pricing power—for example, as a result of imperfections in the market. Over the period from 1976 to 2003, pricing in European industrial sectors has, according to Fitz Gerald *et al.* (Chapter 3, this volume), fluctuated more with European market prices and less with world market prices. The industrial sector most vulnerable to global price competition has been basic metals, but chemicals in The Netherlands displays such vulnerability too. For other sectors, where price fluctuations are associated more with European price changes, impacts of carbon-energy taxes introduced at the EU level would be likely to be moderate. Finally, a few sectors are influenced more by the domestic price index, reflecting either some pricing power in the international market or that goods are traded mainly domestically and with limited international competition; for example, because of unfavourable value-to-weight ratios. When modelling the impacts of carbon-energy taxes, therefore, it is naturally important to take account of such considerations before making assumptions that all cost increases will lead to changes in demand and to resultant loss of market share for affected industries. It is one of the virtues of the ex-post approach to study of carbon-energy taxation that such relations will be captured by historical data and time series for demand and trade.

Winning market share is traditionally regarded as a solid indicator of improved competitiveness. Still, where the global market share increases disproportionately in some regions of the world (say, China), it will mathematically reduce the relative market share of industries in other regions, although absolute sales in those regions are not necessarily affected. Salmons and Miltner (Chapter 4, this volume) consider four other indicators of competitiveness in view of their theoretical properties. Caution is advised when using changes in *sector profitability* as an indicator, because the underlying movements can be difficult to disentangle and interpret. More promising indicators are *export intensity* and *net import penetration,* while *global market share* as an indicator of competitiveness needs careful analysis and interpretation. At the practical level, gathering data on global market shares poses practical difficulties too. If data for European market shares are used instead, these problems are compounded, as the EU25 account for 70–85 per cent of trade for most of the relevant sectors (with the notable exception of chemicals and pharmaceuticals). *Export intensity,* measured as the ratio of a country's exports to its total outputs, is also a reasonable indicator; although when exports account for more than 90 per cent of total output, a threshold for its validity is reached. Net import penetration is the most straightforward indicator in terms of data availability, as it refers to the ratio of a country's net imports of goods and services relative to its domestic demand.

For three indicators, Salmons and Miltner inspect trends for 56 industrial sectors of the seven EU ETR countries. Overall, data availability was not complete for all countries and sectors (with notable difficulties in the case of Finland and Slovenia). While no general signs of loss of competitiveness for the period 1990–2002 can be identified, certain countries and sectors do seem to have performed more convincingly than others. While there were signs of loss of competitiveness for 13 sectors, competitiveness in nine sectors seems to have improved, while the remaining 34 (of 56) display relatively stable performance. Of 13 sectors displaying weaker competitive performance, eight are located in the UK and Germany, where carbon-energy taxation was introduced only towards the end of the decade studied. Both Sweden and Denmark, among the pioneers in environmental tax reform, enjoyed more gains in competitiveness than they suffered losses.

Unit energy costs as a novel indicator of the specific sensitivity to carbon-energy taxation has been explored specifically in COMETR in order to measure the extent to which competitiveness has been affected by carbon-energy taxation. According to Leibenstein's concept of X-efficiency, productivity improvements in the use of energy would become available and could be drawn on as the tax burden increases. This would lead us to expect that unit energy costs as a share of value added would first increase in response to a tax increase, and then level off. The question remains whether we should expect companies to do any more than offset the increased tax burden. In this respect, Porter's hypothesis would suggest that these productivity improvements would lead to innovation of processes and products that could lead to gains in competitive strength in the longer run.

Enevoldsen *et al.* (Chapter 5, this volume) distinguish the potential long-term impacts of less obvious, 'hidden' Porter effects on innovation and demand from more immediate impacts on rising unit energy costs resulting from increased energy prices and taxes. Results from the analysis of the 56 industrial sectors in the seven ETR countries for the years 1990–2003 are striking in the way that increases in energy prices and carbon-energy taxes differ. The long-run impact of an increase in the factor costs of energy is a 0.77 per cent increase in unit energy costs for a 1 per cent increase in energy prices, whereas conversely the impact of a similar increase in energy taxes is just a 0.03 per cent increase in unit energy costs.

Two oil crises have shown that rising costs of energy can have fairly damaging impacts on economies; not surprisingly, as energy is an important factor of production next to capital and labour. The somewhat unexpected difference between the impacts of energy prices and of energy taxes on unit energy costs found over the period studied here, in which energy prices were relatively stable, can perhaps be explained in the following way.

First, whereas rising energy prices will affect relative prices on the world market, taxes are domestic and will only affect final energy use in the country itself. With increases in global energy prices, prices of raw materials will increase too, which will reduce output. In the case of domestic taxes, imported raw materials will not be affected, and less effort will be required to maintain unit energy costs as a share of value added in manufacturing industries. Secondly, energy price increases on world markets provide no revenues to energy-consuming countries that can be recycled to lower other distorting taxes for employers, such as social security contributions. Although these taxes are not directly included when value added is accounted for, they do have distorting properties, and put a brake on economic activity and the ability of manufacturers to adjust their activities to costs of energy that increase. In the case of revenue-neutral environmental tax reform, carbon-energy taxes substitute for other distorting taxes.

Thirdly, it cannot be ruled out that a tax has a different psychological impact than that of a price increase. Market energy prices fluctuate constantly, but an energy tax presents an increase unlikely to disappear again. Company managers may react more strategically to offset negative impacts from a tax, whereas a wait-and-see attitude may prevail in the case of an energy price increase, at least for a while.

As a possible result of mechanisms such as those described above, Enevoldsen *et al.* find only negligible impacts from environmental tax reforms. In the seven European ETR countries, a 10 per cent increase in energy taxes only resulted in a 0.3 per cent increase in unit energy costs. The associated indirect impacts of rising energy taxes on the value of wages, which Bovenberg and de Mooij (1994) warn against, amounts to a small 0.023 per cent increase in unit labour costs for a 10 per cent increase in energy taxes. Together, these two impacts amount to a mere 0.04 per cent decline in gross value added (for a sector with 10% energy costs and 50% labour costs) for a 10 per cent increase in energy taxes. In considering the potential Porter effects, the analysis by Enevoldsen *et al.* finds that the existence of a wider economic impact sufficient to offset the decline in output could not be rejected. A statistically significant relationship is identified between increases in carbon-energy taxes and gross value added for the 56 manufacturing sectors—on average a 10 per cent increase in energy taxes is seen to have led to small increases in gross value added of some 0.23 per cent. This result from an ex-post panel regression analysis is very much in line with the findings of the E3ME model, which also finds small but positive overall impacts arising from environmental tax reform. Inflationary effects from carbon-energy taxes are, on the other hand, not confirmed, and suggest the possibility of reaping a double dividend from environmental tax reform, by extending the architecture of carbon-energy taxation under the principle of revenue neutrality.

Under the present system of carbon-energy taxation, very considerable exemptions have been allowed to energy-intensive industries in Europe.[3] Some sectors are directly exempted under the EU's Energy Taxation Directive, whereas other sectors enjoy domestic exemptions from higher rates of carbon-energy taxes. These industries are generally not labour intensive and so are penalized by a tax shift from labour to energy in the absence of special tax arrangements. Andersen and Speck (Chapter 6, this volume) come to the conclusion that the exemptions have clearly safeguarded competitiveness, in that the tax burden remains at or below 5 per cent of gross operating surplus. In countries where there has been recycling of revenue to lower employers' social security contributions (SSC), energy-intensive industry has seen its carbon-energy tax bill reduced by an amount equivalent to a net 1–2 per cent of gross operating surplus. When the value of the energy productivity improvements induced by the tax (cf. Enevoldsen *et al.*, Chapter 5, this volume) is offset against the tax bill, for less energy-intensive sectors (such as pharmaceuticals and meat processing), a net gain from the shift to carbon-energy taxation is in fact apparent. Such gains are not apparent for the conventional energy-intensive sectors—cement and ferrous and non-ferrous metals.

The coefficient for tax-induced energy savings identified remains a conservative estimate, as cost savings from improvements in energy

---

[3] Under the new environmental guidelines regarding state aid, it has become a legal requirement that member states present a detailed analysis of the ability to pass on the tax in different sectors, as well as providing details for tax burdens as a share of gross value added. Without such information, the European Commission will no longer be able to allow or extend exemption schemes (cf. European Commission, 2008; see also Andersen and Speck, 2009 for a discussion of exemptions and state aid rules).

productivity have taken place that, for most sectors, exceed extra costs from the tax. In some cases, improvements may have taken place as a result of supporting government mechanisms, such as voluntary agreements and revenue recycled in the form of subsidies for technology upgrades, as the findings of Bjørner and Togeby (1999) also indicate. Such energy productivity improvements are apparent especially for Denmark and Germany, where the burden of carbon-energy taxation in less energy-intensive sectors is dwarfed by improved energy productivity. Also, non-ferrous metals display a net gain, whereas for steel and cement, which appear to be the sectors most sensitive to carbon-energy taxation, the results are mixed.

More detailed analyses of steel and cement indicate that the very low levels of carbon-energy tax rates have led to only moderate changes in energy use, and these mainly in terms of fuel shifts. In the steel sector, shifting the technology itself rather than fuels only, by moving from conventional basic oxygen (BOF) to electric arc furnaces, remains a potential option in relation to well over half of the production capacity in Europe. Also, if more scrap iron could replace pig iron, it seems that energy consumption could be reduced by as much as 75 per cent (reflected in the difference between the 19 GJ/1,000 EUR output in Germany and the corresponding 5 GJ/1,000 EUR output in Denmark). For the cement sector, energy consumption is more uniform, approximately 40–50 GJ/1,000 EUR output, and the most efficient technology, dry rotary kilns, is already widely applied in Europe. In Danish cement, new fuels, such as municipal waste, have been brought in to substitute for coal and the sector here in fact counts among the net winners in relation to carbon-energy taxation. Nevertheless, cement accounts for approximately 10 per cent of final energy use in industry and process-related CO_2 emissions are very considerable—the gains associated with identifying substitutes for cement in building and construction activities are therefore likely to be substantial once CO_2 is accorded a price.

With respect to revenue recycling, Sweden represents an interesting case. Experience differs from that of Denmark and Germany; first, because Sweden had already introduced energy taxes on industry in the 1970s and secondly, because Sweden decided to lower income taxes for employees, rather than to reduce employers' labour taxes, when recycling the revenue of its environmental tax reforms. In 1993, Sweden decided to reduce its carbon-energy taxes for industry considerably, which induced a negative trend, whereby energy consumption increased per unit of value added in steel and cement. Unsurprisingly, this confirms that tax rates matter.

Still, the policy of reducing income taxes rather than employers' SSC may have added to the difficulties experienced in Sweden, where sectors faced higher burdens than in Germany and Denmark, as discussed above.

The analysis presented here suggests that short-run competitiveness concerns definitely need to be balanced by a more long-term view on gains from improved energy productivity. Some of these improvements take place as a result of business-as-usual technology upgrades; further improvements are, however, induced by the economic incentive posed by carbon-energy taxes (or allowance prices).

10.5 The environmental dimension of the competitiveness debate: carbon leakage

Carbon leakage is usually defined in quantitative terms as the ratio between non-abating countries' increase in CO_2 emissions and abating countries' reduction in CO_2 emissions. Risk of carbon leakage, in particular relating to energy-intensive industries, has featured prominently among concerns related to unilateral introduction of carbon-energy taxes (as well as to the introduction of auctioned emission allowances in trading schemes, as noted above). There can be no double dividend if industries simply relocate their production to countries without regulatory controls. Even revenue-neutral environmental tax reforms impose a net burden on energy-intensive industries, as their labour force is too small to allow full compensation of carbon-energy taxation via reduced labour costs, as has been clear throughout this book. Because carbon emissions tend to be concentrated in certain energy-intensive industries, a focus on sectoral competitiveness is required for analysis of the carbon leakage dilemma.

Using the E3ME model, Barker *et al.* (Chapter 8, this volume) explore impacts of environmental tax reforms in Europe for carbon leakage more specifically and within the same macroeconomic framework as described above. The mechanisms through which increases in carbon-energy taxes are assumed to impact product prices, in particular those of energy-intensive goods, are the same as in previous chapters. E3ME includes a technological progress indicator (Barker, 1999), reflecting a technological response to improve energy efficiency that in addition serves to improve product quality. The strength of technological progress depends on the level of investments; while the equations for the willingness of customers to pay for improved quality of goods in turn affect both net exports and output. Modelling with E3ME shows a reduction of emissions of greenhouse gases from environmental tax reforms in six member states in the range of 1.5 per cent to 5.9 per cent in 2004—and a reduction for the EU25 as a whole of approximately 1.5 per cent. No other member states increase their emissions in the short run, that is, there is no carbon leakage within Europe. A further E3ME projection for the year 2012 indicates that impacts for the EU25 stabilize at a 1.3 per cent greenhouse gas reduction—approximately 60 million tonne CO_2-equivalent. On the other hand, increases in greenhouse gases for non-ETR member states sum to approximately 1 million tonne CO_2-equivalent. The implication is a carbon leakage rate of approximately 2 per cent.

Although low, the identified leakage rate is not inconsistent with other findings in the literature. With carbon prices around 10 EUR/tonne CO_2, leakage rates of 25–40 per cent as a result of relocation from Annex I to non-Annex I Kyoto countries have been suggested by detailed models of the steel industry, the most sensitive sector (Reinaud, 2004). Taking into account exemptions, the effective carbon-energy tax rate for energy-intensive industries is at present closer to 1–2 EUR/tonne CO_2 for countries with environmental tax reforms. A simple scaling of the results mentioned indicates that we might expect these low tax effective rates to yield leakage at rates of 2.5–4 per cent—and lower for less sensitive industries—well in accordance with the recorded 2 per cent leakage rate arrived at in E3ME. As difficulties in modelling the impacts of carbon-energy taxes on leakage are not small, such scaling is merely illustrative. Still, even some of the more comprehensive CGE models rely only on data for a single year (cf. Barker and Ekins, 2004, and Barker *et al.*, Chapter 8, this volume, for a discussion of modelling deficits).

The analysis of pricing power by Fitz Gerald *et al.* (Chapter 3, this volume) pointed to basic metal industries as the energy-intensive sector most vulnerable to price competition. A simple inspection of production trends shows an increase in crude steel production in Germany, Sweden, Finland, Slovenia, and Netherlands between 1995 and 2004 (Speck and Salmons, 2007). Only the UK experienced a marked decline; this occurred well in advance of the introduction of its climate change levy. Still, relative market shares in the basic metal industries of the ETR countries have declined, but mainly because of a doubling of crude steel production in China, an increase that is largely consumed within China itself.

In the present situation, with rather low effective carbon-energy tax rates for energy-intensive industries, leakage does not appear to warrant serious concern. In fact, exemptions were introduced precisely to avoid leakage and maintain an approximately level playing field with industries in non-abating countries. The preoccupation is more with the levels of leakage rates that could be expected as the carbon price is gradually increased, as a result either of taxes or of full auctioning of allowances under the EU ETS system. As the Swedish experience of the early 1990s indicates, the implications of unilaterally introducing significant carbon-energy taxes can be far-reaching, in particular if coinciding with an economic recession.

10.6 Coping with the dual challenge of GHG stabilization and international competitiveness

Results of modelling studies reviewed by the IPCC (2007a: 659) indicate that the price of carbon will need to be increased globally to a level of 20–50 USD/tonne CO_2 (15–40 EUR/tonne CO_2) by 2030 in order to achieve stabilization of greenhouse gases at a level sufficient to stabilize atmospheric CO_2-equivalent concentrations at 450–550 ppm and presumably avoid dangerous anthropogenic interference with the climate system. There are few modelling studies that address the more desirable stabilization level of 450 ppm, with its lower associated risk of exceeding the 2-degree C target. The relatively limited behavioural change induced by nominal carbon-energy tax rates up to approximately 25 EUR/tonne CO_2 evident in a number of the EU ETR countries lends support to the view that, in order to achieve the reduction in global emissions required by 2030 to comply with stabilization at 450 ppm, the global carbon price will most likely have to reach the upper bound of the results reviewed by IPCC, that is, approximately 40 EUR/tonne CO_2. On the positive side, it can be noted that there is apparently sufficient mitigation potential below a cost of 40 EUR/tonne CO_2 to achieve the stabilization trajectory implied by the Meinshausen figure above (Figure 10.1). This would also conform with the more ambitious 30 per cent reduction by 2020 target that will be aimed for by the European Union if an international agreement on reducing CO_2 emissions is reached.

With the price of carbon in Phase 2 of the EU ETS in the range of 10–25 EUR/tonne CO_2, reaching the 40 EUR/tonne CO_2 level will require a steady annual increase in the carbon price of 1–2 EUR/tonne CO_2

between now and 2030. As the IPCC (2007a: 659) has noted: 'A carbon price that rises over time is a natural feature of an efficient trajectory towards stabilization'. As noted by Ekins (Chapter 9, this volume), the volatility of the EU ETS carbon price could be reduced if an EU-wide carbon tax were introduced to underpin the price mechanism of the trading scheme.

With annual emissions in the EU27 presently at 5 gigatonne of CO_2-equivalents, an increase in the CO_2 price across the whole economy to 40 EUR/tonne CO_2, which curbs emissions by, say, 50 per cent would yield an additional annual tax stream of 100 billion EUR in the European Union (although this would not be net extra revenue, because some of the business tax payments would be set against other tax liabilities). This is revenue that would be available for reducing labour and other taxes if the increase of the carbon price was presented as part of an overall environmental tax reform. In comparison, the switch to more auctioning of carbon allowances from 2013, as proposed by the European Commission, and at foreseen allowance prices, would be expected to yield an annual revenue of 30 to 50 billion EUR (again depending on tax offsets elsewhere). This would also constitute a significant cash sum that could be used to lower other taxes.

While the tax reductions that could be facilitated on the basis of these proceeds would differ among member states according to per capita carbon emissions, an implied revenue of less than 1 per cent of GDP would not allow for sizeable reductions in labour taxes for employers and employees. More significant tax shifts would require that not only carbon but also energy consumption per se became subject to taxation. In this context, it may be recalled that in the European Commission's 1990 proposal for a carbon-energy tax, the tax base was split 50/50 between carbon and energy, a *modus operandi* (as noted by Ekins, Chapter 9, this volume) that is receiving attention again in the European Commission in the current revision of the Energy Taxation Directive (European Commission, 2007a: 7).

Most of the modelling studies reviewed by the IPCC emphasize the availability of carbon emission reductions via savings in energy end-use in buildings and to a lesser extent in industry and transport. The IPCC predicts that such reductions will come to dominate in the first quarter of the present century because of the relatively moderate cost involved. In the second quarter of this century, de-carbonization of energy supply begins to dominate over energy savings as a result of the application of a mix of strategies, including carbon capture and storage (CCS) and low-carbon energy sources. Inertia in the capital stock of power supply and industry is an important factor in understanding the time lags in the transition process to a low-carbon economy. It will take quite some time before the price signals attached to carbon, whether by using taxes or allowances, work their way through the economy.

Obviously, in the meantime, carbon leakage and competitiveness remain a concern for decision-makers. Findings from the COMETR project, as reported here, allow us to address the justification for these concerns from the experience of the tax shift of some 25 billion EUR that took place in some European countries in the 1990s. The shift did not produce carbon leakage at any grave rate. Nevertheless, the problem of carbon leakage is likely to be perceived in relation to basic trends in globalization as well as in relation to the current fiscal crisis and recession as it unfolds.

We noted above that China has doubled its steel production over the past decade and the same is true for cement. China now is responsible for 48 per cent of global cement production. It is process-related emissions from cement that in 2007 made China surpass the USA as the leading global emitter of greenhouse gases. China, in only eight years, from 2000–7 has doubled its greenhouse gas emissions and now accounts for 6 gigatonne of CO_2 annually or 20 per cent of global emissions. While about a third of China's emissions can be related to the country's exports, growth in energy-intensive industries, such as steel and cement, is very much a response to increases in domestic demand. Whereas only 20 per cent of the Chinese population resided in cities before the economic revolution, the share of urban dwellers is growing at a rapid rate, and in less than 15 years has exceeded 40 per cent. Chinese cities are mushrooming, with urban structures and high-rise buildings, to accommodate the transition to an industrialized society; and similar transitions are evident in other Asian countries too. Steel and cement is in demand in Asia, and market growth has taken place with few implications for production in European and other Western countries. While China has had a reputation for inefficient planned-economy-type production facilities, the recent growth has facilitated uptake of new energy-efficient technologies, to an extent that energy consumption per tonne of cement, according to IEA figures, is now slightly lower in China than in the USA (IEA, 2007a: 161). While carbon leakage poses a problem for competitiveness, it is worth recalling that it only exacerbates the environmental problem to the extent that carbon intensities are higher in the regions where leakage manifests itself. China's strong emphasis on improving energy efficiency and its investment in advanced technology now call into question the extent to which this is still the case.

Carbon-energy taxation is on the agenda in many countries, including China and other industrializing countries. In the past, the possibility for border-tax adjustments to carbon-energy taxes, applied unilaterally or regionally, preoccupied a great deal of the economic and legal debate in Europe. A border-tax arrangement in relation to a carbon-energy tax implies, from the perspective of the country that imposes a carbon-energy tax, that while all imports are levied with a customs fee corresponding to the carbon-energy content, all exports are conversely offered a refund on a similar basis. Because of international agreements under the World Trade Organization (WTO), border-tax arrangements may be problematic. WTO rules strictly prohibit the introduction of new custom taxes that relate only to imports, but may offer some leeway for process-related taxes in relation to environmental burdens. The legal and economic difficulties related to WTO regulations can be avoided if emerging industrializing countries were alternatively to impose a carbon-energy tax on their exports to the abating countries. This option has surfaced in debates on competitiveness impacts associated with carbon-pricing schemes and possible border-tax adjustments. In countries such as China, consideration of such measures is driven by a concern to avoid border-tax arrangements in Annex I countries that could be used to penalize exports from emerging countries and impose additional barriers to trade.

A carbon-energy tax on products exported from developing countries into Annex I countries, but imposed by the exporting, developing nation, could in fact help reduce concern about carbon leakage, while at the same time generating a significant revenue flow to developing countries.[4] Such arrangements would need to remain a purely transitional measure to more widespread carbon-energy pricing, including in developing countries, and would clearly be helpful in the medium term in order to achieve stabilization of atmospheric carbon concentrations. China is aiming to quadruple its GDP per capita over the next two decades, but even under a scenario with improved energy productivity, its energy consumption and the related CO_2 emissions are likely to double from the present level

---

[4] The idea of a carbon export tax was presented informally by Chinese members of the Task Force on Economic Instruments and Energy Efficiency under the China Council for International Cooperation on Environment and Development (CCICED) in April 2008.

of 6 gigatonnes to about 12 gigatonnes. Accelerated energy productivity could at best reduce annual emissions to 9 gigatonnes (IEA, 2007b: 370). Now a net importer of energy, China is facing an energy supply crisis that it shares with the rest of the world, and its domestic reserves of coal are no longer what they were once believed to be. As fossil fuel reserves dwindle, energy supply is seen more and more as the bottleneck preventing improvement in material standards of life for Asia's populous nations. Under these perspectives, carbon-energy taxation that is phased in by developing countries, first on exports to Annex I countries and next, with lower rates adjusted to domestic price levels, on energy consumption, would offer a promising approach to realign climate policy with concerns about leakage and equity. It should be noted that taxation is the only option considered by China and other developing countries—there is no intention on the part of these countries to participate in a global carbon trading system that would cap their emissions. The negotiating position of many developing nations is based on the belief that the Annex I countries need to accept higher relative burdens due to the historical responsibility they bear for the accumulated stock of greenhouse gases in the atmosphere.

The findings of the COMETR project have improved our general understanding of the opportunities and mechanisms associated with carbon-energy taxation, in particular when introduced as part of revenue-neutral environmental tax shifts. While much of the ex-ante modelling has been speculative in nature, the ex-post approach applied in COMETR is constrained by the historical data. Different analytical techniques and approaches have been applied to explore the experiences with carbon-energy taxation in seven European countries and the various findings are relatively consistent in showing the absence of strongly negative impacts on the economy, while at the same time indicating a plausible, small double dividend. The overall tax shift of around 25 billion EUR is by no means trivial in relation to the carbon tax revenue of around 100 billion EUR for the EU27 implied in the IPCC recommendation of a 40 EUR/tonne CO_2 tax by 2030. Although tax-induced price increases are politically manipulated, the results from COMETR seem to indicate that reactions from target groups to these price signals are more prompt and decisive than those in response to changes in fluctuating world market energy prices. While this observation seems to warrant further research, it does underline the possible strength of a policy committing to a carbon-energy price escalator as proposed by the IPCC. Concerns about carbon leakage from the abating countries will be the main environmental argument against such a policy, but such concerns are sometimes overstated. Moreover, the brief review of the wider economic dynamics in China and other developing economies seems to indicate that a more phased and global approach to carbon-energy pricing could be developed to effectively tackle carbon leakage. The introduction of earlier carbon-energy taxation in Europe against the background of this evolving global context certainly deserves further analytical scrutiny in the future.

References

Agnolucci, P., Barker, T., and Ekins, P. 2004. 'Hysteresis and energy demand: the announcement effects and the effects of the UK climate change levy'. Tyndall Centre for Climate Change Research. Working Paper 51. Norwich: University of East Anglia.

Andersen, M. S, Dengsøe, N., and Pedersen, A. B. 2001. 'An evaluation of the impact of green taxes in the Nordic countries'. TemaNord 566, Nordic Council of Ministers, Copenhagen.

——and Speck, S. 2009. 'Environmental tax reforms in Europe: mitigation, compensation and CO_2-stabilisation', in J. Cottrell, J. E. Milne, H. Ashiabor, L. Kreiser, and K. Deketelaere (eds.), *Critical Issues in Environmental Taxation: International and Comparative Perspectives, Volume VI*. Oxford: Oxford University Press.

Barker, T. 1999. 'Achieving a 10 % cut in Europe's carbon dioxide emissions using additional excise duties: coordinated, uncoordinated and unilateral action using the econometric model E3ME'. *Economic Systems Research*, 11/4: 401–21.

——and Ekins, P. 2004. 'The costs of Kyoto for the USA economy'. *Energy Journal*, 25/3: 53–71.

——Quereshi, M., and Köhler, J. 2006. 'The costs of greenhouse-gas mitigation with induced technological change: a meta-analysis of estimates in the literature'. 4CMR (Cambridge Centre for Climate Change Mitigation Research), Cambridge University, Cambridge.

Bjørner, T. B., and Togeby, M. 1999. 'Industrial companies' demand for energy based on a micro panel database: effects of CO_2 taxation and agreements on energy savings'. ACEE Summer Study 1999. New York: American Council for an Energy-efficient Economy.

Bovenberg, A. L., and de Mooij, R. A. 1994. 'Environmental levies and distortionary taxation'. *American Economic Review*, 84/4: 1085–9.

den Elzen, M. G. J., and Meinshausen, M. 2006. 'Multi-gas emission pathways for meeting the EU 2°C climate target', in J. S. Schellnhuber, W. Cramer, N. Nakicenovic, T. M. L. Wigley, and G. Yohe (eds.), *Avoiding Dangerous Climate Change*. Cambridge: Cambridge University Press.

Enevoldsen, M. 2005. *The Theory of Environmental Agreements and Taxes: CO₂ Policy Performance in Comparative Perspective*. Cheltenham: Edward Elgar.

——Ryelund, A. V., and Andersen, M. S. 2007. 'Decoupling of industrial energy consumption and CO_2-emissions in energy-intensive industries in Scandinavia'. *Energy Economics*, 29/4, 665–92.

European Commission. 2007a. Green Paper on market-based instruments for environment and related policy purposes. COM(2007) 140 final, 28 Mar. Brussels. Available at <http://www.eur-lex.europa.eu/LexUriServ/site/en/com/2007/com2007_0140en01.pdf>.

——2007b. Accompanying document to the proposal for a directive amending Directive 2003/87/EC (climate package impact assessment). Brussels.

——2008. Community guidelines on state aid for environmental protection. OJ C82/01–33, 1 Apr.

IEA. 2007a. *Tracking Industrial Energy Efficiency and CO₂ Emissions*. Paris: IEA.

——2007b. *World Energy Outlook: China and India Insights*. Paris: IEA.

IPCC. 2007a. Working Group III Report Mitigation of Climate Change. Fourth Assessment Report (AR4). Cambridge: Cambridge University Press.

——2007b. Working Group III Report Mitigation of Climate Change. Fourth Assessment Report (AR4). Summary for Policymakers. IPCC. <http://www.ipcc.ch/>.

Knigge, M., and Görlach, B. 2004., Auswirkungen der Ökologischen Steuerreform auf Private Haushalte'. Forschungsprojekt im Auftrag des Umweltbundesamtes. Ecologic, Berlin.

Kononov, S. 2007. 'Kyoto protocol: data, policies, infrastructures'. Presentation. Bonn: UNFCCC press conference. 20 Nov.

Meinshausen, M. 2006. 'What does a 2°C target mean for greenhouse gas concentrations? A brief analysis based on multi-gas emission pathways and several climate sensitivity uncertainty estimates', in J. S. Schellnhuber, W. Cramer, N. Nakicenovic, T. M. L. Wigley, and G. Yohe (eds.); *Avoiding Dangerous Climate Change*. Cambridge: Cambridge University Press.

——Hare, B., Wigley, T. M. L., van Vuuren, D., den Elzen, M. G. J., and Swart, R. 2006. 'Multi-gas emission pathways to meet climate targets'. *Climatic Change*, 75/1: 151–94.

Reinaud, J. 2004. 'Industrial competitiveness under the European Union emissions trading scheme'. Dec. IEA Information Paper. Paris: International Energy Agency.

Richardson, K., Steffen, W., Schellnhuber, H.J., Alcamo, J., Barker, T., Kammen, D., Leemans, R., Liverman, D., Munasinghe, M., Osman-Elasha, B., Stern, N. and Wæver, O. 'Synthesis Report from Climate Change: Global Risks, Challenges

and Decisions, Copenhagen, 10–12 March 2009, http://climatecongress.
ku.dk/pdf/synthesisreport.

Speck, S., and Salmons, R. 2007. 'Leakage analysis within a decoupling framework',
in M. S. Andersen, T. Barker, E. Christie, P. Ekins, J. Fitz Gerald, J. Jilkova,
J. Junankar, M. Landesmann, H. Pollitt, R. Salmons, S. Scott, and S. Speck,
Competitiveness Effects of Environmental Tax Reforms (COMETR). Final report to the
European Commission, DG Research and DG TAXUD. National Environmental
Research Institute, University of Aarhus.

Stern, N. 2007. *The Economics of Climate Change*. Cambridge: Cambridge University
Press.

UNFCCC. 2007. 'Emissions of industrialized countries rose to an all-time high in
2005'. Press release. Bonn.

Annex: Effective Carbon-energy Tax Rates for Industry in Countries with Environmental Tax Reform (by Stefan Speck)

Table A.1. Overview of major tax shifting programmes in EU15

Country	Tax Shift		Comments and remarks	Size of tax shift and revenue neutrality
	From	To		
Denmark 1993	Reduction in tax rates on personal income	Increase in existing taxes on fossil fuels, electricity and waste and new taxes on piped water, wastewater and carrier bags and payroll taxes	The reductions in income taxes were compensated by increase in environmental taxes 12 billion DKK payroll taxes 22 billion DKK and from other measures, such broadening the tax base (11 billion DKK)	Revenue neutral in 1998—but not in the first years of the reform; Total shift in 1998: 3.9% of GDP and increase in environmental taxes: 1% of GDP
Denmark 1995	Reduction in the social security contributions, supplementary pension payments and investment subsidies for energy savings	Increase in energy taxes (but industry is reimbursed when entering voluntary agreements) and new tax on SO_2 and natural gas	The major part of revenues generated from the energy taxes were planned for funding investments of energy-savings measures in enterprises, to reduce employers' social security contributions and for support of small- and medium-sized enterprises	Revenue neutral: 2000—planned revenues from environmental taxes: 0.2% of GDP
Denmark 1998	Reduction in tax rates on personal income for lower and middle incomes	Increase in energy taxes (by 15–25 per cent) and property tax		No revenue neutrality per annum: see below table

Finland 1997	Personal income (state and local) and employers' social security contributions	Energy, CO_2 tax and landfill tax	Reduction of tax revenues of 5.5 billion FIM was planned to be partly financed by revenues from the energy tax (1.1 billion FIM) and from higher landfill tax rates (300 million FIM) in 1997.	No revenue neutral reform reduction: 0.9% of GDP increase: 0.2% of GDP
Finland 1998	Labour taxation	Energy and environmental taxes plus corporate profit tax (broadening of tax base)	Planned reduction in revenues generated from labour taxation: 1998—1.5 billion FIM 1999—3.5 billion FIM	No revenue neutral reform—planned deficit of 1.5 billion FIM in 1998 and 2.5 billion FIM in 1999 reduction 1999: 0.5% of GDP increase 1999: 0.14% of GDP
Germany 1999–2003 (a five year programme)	Employers and employees social security contributions	Energy (mineral oils, natural gas and electricity)	A reduction of around 1.7% of employers' and employee's pension contribution in 2003; revenues from energy taxes amounted to 18.6 billion EUR in 2003	Revenue neutrality is part of the programme—1 bill EUR is earmarked for budget consolidation[1]. Revenue 2003: 18.7 billion EUR—0.9% of GDP
The Netherlands 1996, (further increases in the regulatory energy tax in 1999, 2000, and 2001 generating additional 1.5 billion EUR—part of a complete overhaul of the fiscal framework)	Personal income, corporate profits, employers' social security contributions	Energy and CO_2 (regulatory energy tax), water, waste disposal	Revenues were planned to be recycled back (1996) <ul><li>by cutting employers' social security contributions by 0.19%;</li><li>by raising the tax credit for self-employed people by 1300 Dfl;</li><li>by reducing the corporate income tax by 3% for the first 100,000 Dfl of profits;</li><li>by reducing the income tax rate by 0.6%; and</li><li>by raising the standard income free allowance by 80 Dfl and the tax free allowance for older people by 100 Dfl.</li></ul> Around 930 million Dfl were planned to be recycled to industry and 1,230 million Dfl to households.	Revenue neutral reform

(cont.)

Table A.1. (Continued)

Country	Tax Shift		Comments and remarks	Size of tax shift and revenue neutrality
	From	To		
Sweden 1991	Personal income tax and social security contributions	Environmental and energy taxes including CO_2 tax and SO_2 tax as well as VAT on energy products	ETR was part of a major fiscal reform—personal income taxes were planned to be cut by 71 billion SEK and environmental taxes were planned to be increased by 18 billion SEK	No revenue neutral reform
Sweden—10 year programme: 2001–2010	Personal income tax and personal social security contributions	Environmental taxes	A total of 30 billion SKR (3.3 billion EUR) shall be shifted at the end of the ten year programme in 2010 (corresponding to almost 1.4% of GDP)—Green tax shift programme is part of a much bigger reform	No revenue neutral reform: until 2003 (the first three years): 8 billion SEK additional generated from environment taxes, eg. around 0.3% of GDP
UK 1996	Employers' national insurance contribution (NIC)	Landfill levy	Revenues are used for a reduction of 0.2% in employers' NIC from 10.2% to 10%	Budget 2004 (forecast) 0.6 billion UK£—0.05% of GDP
UK 2001	Employers' national insurance contribution (NIC)	Energy/CO_2 emissions under the Climate Change Levy (CCL)	Revenues are used for a reduction of 0.3% in employers' NICs; revenue is estimated to be around 1 billion UK£ pa	Budget 2004 (forecast) 0.8 billion UK£—0.06% of GDP
UK 2002	Employers' national insurance contribution (NIC)	Aggregates levy (sand, gravel, crushed rock)	Revenues are used for a reduction of 0.1% in employers' NICs; revenue estimated at ~305 million UK£ in 2002/3	Budget 2004 (forecast) 0.3 billion UK£—0.02% of GDP

[1] A slight discrepancy regarding the usage of ETR revenues can be found in different publications of the German Government. It is mentioned in a report published by the Ministry of Environment in 2004 that around 1 billion EUR is used for budget consolidation. In slight contrast to this statement the Ministry of Finance argues in a press notice in September 2005 that the revenues, which are not hypothecated for reducing pension contributions, are being used for the supporting of renewable energy sources and in addition to compensate the shortfall of tax revenues as a consequence of the financial provision given to all the biogenic fuels. These fuels are exempt from mineral-oil taxes.

Source: EEA 2005 (forthcoming) and author's data compilation.

Annex A.2 to A.18: Development of Energy/CO_2 taxes in selected EU member states[2]

Table A.2. Total energy taxes in Denmark

	Petrol unleaded DKK/ 1,000l	Diesel (transport) DKK/ 1,000l	Light fuel oil DKK/ 1,000l	Heavy fuel oil DKK/ ton	Coal DKK/ ton	Natural gas DKK/ 1,000 m³	Electricity-other purposes DKK/ MWh
1988	3,330	1,760	1,760	1,980	675		326
1989	3,330	1,760	1,760	1,980	765		328.6
1990	2,250	1,760	1,760	1,980	765		330
1991	2,250	1,760	1,760	1,980	765		330
1992	2,250	1,760	1,760	1,980	932		370
1993	2,250	1,760	1,760	1,980	932		370
1994	2,450	2,040	1,760	1,980	932		400
1995	3,020	2,270	1,760	1,980	1,012		430
1996	3,270	2,290	1,760	1,980	1,102	230	460
1997	3,320	2,390	1,760	1,980	1,192	1,450	500
1998	3,370	2,390	1,970	1,980	1,282	1,690	566
1999	3,770	2,620	1,970	2,230	1,492	1,690	581
2000	3,870	2,850	2,000	2,270	1,542	1,820	636
2001	3,970	3,030	2,050	2,320	1,592	2,180	651
2002	4,070	3,030	2,100	2,380	1,667	2,240	666
2003	4,070	3,030	2,100	2,380	1,667	2,240	666
2004	4,070	3,030	2,100	2,380	1,667	2,240	666
2005	4,070	3,030	2,100	2,380	1,667	2,240	666
2006	4,070	3,030	2,100	2,380	1,667	2,240	666
2007	4,070	3,030	2,100	2,380	1,667	2,240	666

Note: The Danish SO_2 tax is not considered in the tables.

[2] An attempt was made to present all tax rates valid as of January 1st of the relevant years. However, we depart from this procedure if a major tax revision took place during a specific year meaning that the tax rates shown in the following tables are not the ones which were valid on January 1st. For example, the German eco-tax reform was implemented on April 1st 1999 and therefore we are showing the tax rates valid on April 1st 1999 for the year 1999. In addition, the UK tax rates were calculated as annual average rates.

Data have been compiled by the author and colleagues working at partner institutes of the COMETR project using a whole range of different reports published by institutions such as national statistical offices, international organizations, etc.

Table A.3. Energy taxes in Denmark

	Petrol unleaded DKK/ 1,000l	Diesel (transport) DKK/ 1,000l	Light fuel oil DKK/ 1,000l	Heavy fuel oil DKK/ ton	Coal DKK/ ton	Natural gas DKK/ 1,000 m³	Electricity-other purposes DKK/ MWh
1988	3,330	1,760	1,760	1,980	675		326
1989	3,330	1,760	1,760	1,980	765		328.6
1990	2,250	1,760	1,760	1,980	765		330
1991	2,250	1,760	1,760	1,980	765		330
1992	2,250	1,760	1,760	1,980	932		370
1993	2,250	1,490	1,490	1,660	690		270
1994	2,450	1,770	1,490	1,660	690		300
1995	3,020	2,000	1,490	1,660	770		330
1996	3,270	2,020	1,490	1,660	860	10	360
1997	3,320	2,120	1,490	1,660	950	1,230	400
1998	3,370	2,120	1,700	1,660	1,040	1,470	466
1999	3,770	2,350	1,700	1,910	1,250	1,470	481
2000	3,870	2,580	1,730	1,950	1,300	1,600	536
2001	3,970	2,760	1,780	2,000	1,350	1,960	551
2002	4,070	2,760	1,830	2,060	1,425	2,020	566
2003	4,070	2,760	1,830	2,060	1,425	2,020	566
2004	4,070	2,760	1,830	2,060	1,425	2,020	566
2005	3,850	2,820	1,860	2,090	1,445	2,040	576
2006	3,850	2,820	1,860	2,090	1,445	2,040	576
2007	3,850	2,820	1,860	2,090	1,445	2,040	576

Table A.4. CO$_2$ taxes in Denmark

	Petrol unleaded DKK/ 1,000l	Diesel (transport) DKK/ 1,000l	Light fuel oil DKK/ 1,000l	Heavy fuel oil DKK/ ton	Coal DKK/ ton	Natural gas DKK/ 1,000 m³	Electricity-other purposes DKK/ MWh
1992							
1993		270	270	320	242		100
1994		270	270	320	242		100
1995		270	270	320	242		100
1996		270	270	320	242	220	100
1997		270	270	320	242	220	100
1998		270	270	320	242	220	100
1999		270	270	320	242	220	100
2000		270	270	320	242	220	100
2001		270	270	320	242	220	100
2002		270	270	320	242	220	100
2003		270	270	320	242	220	100
2004		270	270	320	242	220	100
2005	220	240	240	290	222	200	90
2006	220	240	240	290	222	200	90
2007	220	240	240	290	222	200	90

Note: CO$_2$ tax data are only presented in 1993 although the CO$_2$ tax rate was already introduced in May 1992.

Table A.5. Effective CO_2 tax rate for businesses

	Light process CO_2 tax rate		Heavy process CO_2 tax rate	
	With agreement DKK/ton CO_2	**Without agreement DKK/ton CO_2**	**With agreement DKK/ton CO_2**	**Without agreement DKK/ton CO_2**
1992				
1993		50		5
1994		50		5
1995		50		5
1996	50	50	3	5
1997	50	60	3	10
1998	50	70	3	15
1999	58	80	3	20
2000	68	90	3	25
2001	68	90	3	25
2002	68	90	3	25
2003	68	90	3	25
2004	68	90	3	25
2005	68	90	3	25
2006	68	90	3	25
2007	68	90	3	25

Note: The nominal CO_2 tax rate was reduced to 90 DKK/ton CO_2 from 100 DKK/ton CO_2 in 2005. However, this reduction was not passed on to businesses, i.e. the effective tax rates have been kept constant in 2005 meaning that the share has been increased that businesses have to pay.

Table A.6. Energy taxes in Finland

	Petrol unleaded EUR/ 1,000l	Diesel (transport) EUR/ 1,000l	Light fuel oil EUR/ 1,000l	Heavy fuel oil EUR/ ton	Coal EUR/ ton	Natural gas EUR/ m³	Households Electricity EUR/ MWh	Industry Electricity EUR/ MWh
1988	146	124	0	0	0	0		
1989	160	142	0	0	0	0.002		
1990	260	127.5	3.4	3.4	2.7	0.002		
1991	316	134.5	3.5	3.5	2.8	0.002		
1992	359	134.5	3.5	3.5	2.8	0.002		
1993	479	150.5	14.1	11.2	5.6	0.004		
1994	401	198	20.5	19.8	11.3	0.011		
1995	452	300	30.2	31.2	19.5	0.009		
1996	519	300	30.2	31.2	19.5	0.009		
1997	519	300	48.8	37.2	28.4	0.012	5.6	2.4
1998	552	326	52.0	44.0	33.4	0.014	5.6	3.4
1999	552	325	63.7	54.0	41.4	0.017	6.9	4.2
2000	552	325	63.7	54.0	41.4	0.017	6.9	4.2
2001	552	325	63.7	54.0	41.4	0.017	6.9	4.2
2002	552	325	63.7	54.0	41.4	0.017	6.9	4.2
2003	581.3	343	67.1	56.8	43.5	0.018	7.2	4.4
2004	581.3	343	67.1	56.8	43.5	0.018	7.2	4.4
2005	581.3	343	67.1	56.8	43.5	0.018	7.2	4.4
2006	581.3	343	67.1	56.8	43.5	0.018	7.3	4.4
2007	587.5	346	70.6	59.6	44.7	0.019	7.4	2.3

Notes: Strategic stockpile fee and oil pollution levy levied on all energy products are not reported. Electricity tax scheme changed in 1997—until 1997 Finland adopted an input tax scheme, i.e. energy products used for electricity generation were subject to energy taxation.

Table A.7. Energy tax rates in Germany

	Petrol unleaded EUR/ kl	Diesel (transport) EUR/ kl	Gas oil EUR/ kl	Heavy fuel oil EUR/ ton	Coal EUR/ ton	Natural gas -used for heating EUR/ MWh	Industry Electricity– effective EUR/ kWh	Households Electricity EUR/ kWh
1988	240	230	8.2	7.2	0	0.0	0.005	0.007
1989	291	230	28.1	14.5	0	1.3	0.006	0.009
1990	291	230	28.4	14.6	0	1.3	0.005	0.009
1991	307	230	39.0	14.6	0	1.8	0.005	0.009
1992	419	280	39.6	14.9	0	1.8	0.005	0.008
1993	419	280	41.2	15.5	0	1.9	0.005	0.009
1994	501	317	41.7	15.6	0	1.9	0.006	0.010
1995	501	317	42.7	16.0	0	1.9	0.006	0.011
1996	501	317	41.9	15.7	0	1.9	0	0
1997	501	317	40.7	15.3	0	1.8	0	0
1998	501	317	40.9	15.2	0	1.8	0	0
1999	532	348	61.4	15.3	0	3.5	0.00205	0.01023
2000	562	378	61.4	15.3	0	3.5	0.00256	0.01278
2001	593	409	61.4	17.9	0	3.5	0.00307	0.01534
2002	624	440	61.4	17.9	0	3.5	0.00704	0.0179
2003	655	471	61.4	25	0	5.5	0.0123	0.0205
2004	655	471	61.4	25	0	5.5	0.0123	0.0205
2005	655	471	61.4	25	0	5.5	0.0123	0.0205
2006	655	471	61.4	25	0	5.5	0.0123	0.0205
2007	655	471	61.4	25	8.4	5.5	0.0123	0.0205

Table A.8. The effects of the German ETR on energy tax rates

		1999	April 1999	January 2000	January 2001	January 2002	January 2003	January 2004
Tax Rates								
Petrol unleaded	EUR/1,000 litres	501.1	531.7	562.4	593.1	623.8	654.5	654.5
Diesel	EUR/1,000 litres	317.0	347.7	378.4	409.0	439.7	470.4	470.4
Natural gas	EUR/MWh	24.3	25.8	27.3	28.8	30.3	31.8	31.8
Natural gas (heating)	EUR/MWh	1.8	3.5	3.5	3.5	3.5	5.5	5.5
LPG	EUR/1,000 kg	952.5	1,005.5	1,058.4	1,111.2	1,164.1	1,217.0	1,217.0
LPG (heating)	EUR/1,000 kg	25.6	38.4	38.4	38.4	38.4	60.6	60.6
Light Fuel Oil (LFO)	EUR/1,000 litres	40.9	61.4	61.4	61.4	61.4	61.4	61.4
Reduced LFO	EUR/1,000 litres	18.4	34.8	34.8	34.8	34.8	34.8	34.8
Heavy fuel oil (HFO)—generation of heat	EUR/1,000 kg	15.3	15.3	17.9	17.9	17.9	25.0	25.0
HFO—generation of electricity	EUR/1,000 kg	28.1	28.1	17.9	17.9	17.9	25.0	25.0
Electricity	EUR/MWh	0.0	10.2	12.8	15.3	17.9	20.5	20.5
Electricity—night storage heating	EUR/MWh	0.0	5.1	6.4	7.7	9.0	12.3	12.3
Electricity—manufacturing industry, agriculture	EUR/MWh	0.0	2.1	2.6	3.1	3.6	12.3	12.3
Increase in tax rates p.a.								
Petrol unleaded	EUR/1,000 litres		30.7	30.7	30.7	30.7	30.7	
Diesel	EUR/1,000 litres		30.7	30.7	30.6	30.7	30.7	
Natural gas	EUR/MWh		1.5	1.5	1.5	1.5	2.9	
Natural gas (heating)	EUR/MWh		1.7	1.7	1.7	1.7	2	
LPG	EUR/1,000 kg		53.0	52.9	52.9	52.9	52.9	
LPG (heating)	EUR/1,000 kg		12.8	12.8	12.8	12.8	22.2	
LFO	EUR/1,000 litres		20.5	0.0	0.0	0.0	0.0	
Reduced LFO	EUR/1,000 litres		16.4	0.0	0.0	0.0	0.0	
HFO—generation of heat	EUR/1,000 kg		0.0	2.6	0.0	0.0	7.1	
HFO—generation of electricity	EUR/1,000 kg		0.0	-10.2	0.0	0.0	7.1	
Electricity	EUR/MWh		10.2	2.6	2.6	2.6	2.6	
Electricity—night storage heating	EUR/MWh		5.1	1.3	1.3	1.3	3.3	
Electricity—manufacturing industry, agriculture	EUR/MWh		2.1	0.5	0.5	4.0	5.3	

Table A.9. Development of energy taxes introduced as part of the ETR

	Petrol unleaded EUR/kl	Diesel (transport) EUR/kl	Gas oil EUR/kl	Heavy fuel oil EUR/ton	Natural gas EUR/MWh	Electricity EUR/kWh
1999	30.7	30.7	20.5	0	1.7	0.01023
2000	61.4	61.4	20.5	2.6	1.7	0.01278
2001	92.1	92.1	20.5	2.6	1.7	0.01534
2002	122.8	122.8	20.5	2.6	1.7	0.0179
2003	153.5	153.5	20.5	7.1	3.7	0.0205
2004	153.5	153.5	20.5	7.1	3.7	0.0205
2005	153.5	153.5	20.5	7.1	3.7	0.0205

Table A.10. Effective tax rates for manufacturing industry, agriculture, forestry and fishery

	Gas oil EUR/kl	Natural gas EUR/MWh	Electricity EUR/kWh
1999	45.0	2.2	0.002046
2000	45.0	2.2	0.002556
2001	45.0	2.2	0.003068
2002	45.0	2.2	0.00358
2003	53.2	4.0	0.0123
2004	53.2	4.0	0.0123
2005	53.2	4.0	0.0123

Note: The tax rates presented are taking into account already existing energy taxes (i.e. in place before 1999) and the energy taxes levied under the ETR. Additional tax regulation known under the term 'Spitzenausgleich' are not considered.

Table A.11. Development of tax rates in The Netherlands

	Petrol unleaded EUR/1000l	Diesel (transport) EUR/1000l	Gas oil EUR/1000l	Light fuel oil EUR/1000l	Heavy fuel oil EUR/ton	Coal EUR/ton
1988	338.6	123.6		52.4	16.8	2.5
1989	337.6	122.2		50.8	17.1	2.5
1990	348.9	174.0		52.9	20.3	5.5
1991	394.4	193.9		53.6	22.5	9.1
1992	443.2	212.3		62.0	29.2	10.9
1993	461.9	272.0		65.0	29.9	10.4
1994	518.1	312.5		66.1	30.2	10.5
1995	527.23	315.83	62.08	61.98	31.72	11.14
1996	528.29	316.35	74.17	73.98	31.11	10.93
1997	514.70	310.29	71.78	71.60	30.11	10.58
1998	565.81	331.58	97.36	96.82	29.99	10.53
1999	581.85	341.01	117.43	116.84	30.52	10.83
2000	591.76	346.81	139.29	138.61	30.77	11.02
2001	602.41	353.03	187.40	186.24	31.04	11.22
2002	621.08	353.26	191.82	196.89	31.53	11.99
2003	643.44	351.34	198.1	196.99	32.11	11.99
2004	658.88	365.77	206.5	205.23	32.51	12.28
2005	668.10	364.91	207.61	206.28	32.51	12.45
2006	668.10	364.91	209.06	207.72	32.51	12.45
2007	678.79	370.75	209.58	208.24	32.51	12.76

Note: Strategic stockpile fee is not included in the tax rates.

Table A.12. Tax levied on natural gas

Annual consumption between	EUR/m³						
	0-800	801-5,000	5,001-170,000	170,001-1 M	1 M-10 M	>10 M-non-business use	>10 M-business use
1988	0.0003	0.0003	0.0003				
1989	0.0003	0.0003	0.0003				
1990	0.0020	0.0020	0.0020				
1991	0.0042	0.0042	0.0042				
1992	0.0111	0.0111	0.0111				
1993	0.0095	0.0095	0.0095				
1994	0.0096	0.0096	0.0096				
1995	0.0103	0.0103	0.0103	0.0103	0.0103	0.0067	0.0067
1996	0.0101	0.0250	0.0250	0.0101	0.0101	0.0065	0.0065
1997	0.0097	0.0387	0.0387	0.0097	0.0097	0.0063	0.0063
1998	0.0097	0.0526	0.0526	0.0097	0.0097	0.0063	0.0063
1999	0.0100	0.0825	0.0825	0.0825	0.0100	0.0065	0.0065
2000	0.0102	0.1046	0.0621	0.0172	0.0102	0.0066	0.0066
2001	0.1306	0.1306	0.0665	0.0208	0.0103	0.0068	0.0068
2002	0.1350	0.1350	0.0689	0.0217	0.0110	0.0070	0.0070
2003	0.1395	0.1395	0.0710	0.0221	0.0110	0.0073	0.0073
2004	0.1429	0.1429	0.0727	0.0113	0.0113	0.0106	0.0075
2005	0.1494	0.1494	0.1019	0.0311	0.0115	0.0107	0.0076
2006	0.151	0.151	0.124	0.034	0.0116	0.0108	0.0077
2007	0.153	0.153	0.134	0.037	0.0118	0.011	0.0077

Table A.13. Tax levied on electricity

Annual consumption between	EUR/kWh					
	0-800	800-10,000	10,000-50,000	50,000-10 M	>10 M-non business use	>10 M-business use
1996	0	0.0138	0.0138	0	0	0
1997	0	0.0133	0.0133	0	0	0
1998	0	0.0133	0.0133	0.000	0	0
1999	0	0.0225	0.0147	0.001	0	0
2000	0	0.0372	0.0161	0.002	0	0
2001	0.0583	0.0583	0.0194	0.0059	0	0
2002	0.0601	0.0601	0.02	0.0061	0	0
2003	0.0639	0.0639	0.0207	0.0063	0	0
2004	0.0654	0.0654	0.0065	0.0065	0.001	0.0005
2005	0.0699	0.0699	0.0263	0.0086	0.001	0.0005
2006	0.0705	0.0705	0.03430	0.0094	0.001	0.0005
2007	0.0716	0.0716	0.03690	0.00102	0.001	0.005

Table A.14. Overview of energy tax development in Slovenia

Year	Petrol unleaded SIT/1000 l	Diesel (transport) SIT/1000 l	Light fuel oil SIT/1000 l	Heavy fuel oil SIT/ton	Natural gas SIT/1000 m³	Coal SIT/ton	Electricity SIT/MWh
1990	140%	90%	20%	32%	5%	5%	5%
1991	140%	90%	20%	32%	5%	5%	5%
1992	140%	90%	20%	32%	5%	5%	5%
1993	140%	90%	20%	32%	5%	5%	5%
1994	140%	90%	20%	32%	5%	5%	5%
1995	140%	90%	20%	32%	5%	5%	5%
1996	140%	90%	20%	32%	5%	5%	5%
1997	140% 2,200	90% 2,600	0 2,600 (2,600)	0 3,100 (3,100)	0 1,300 (1,300)	0 2,500 (2,500)	0
1998	140% 6,600	90% 7,800	0 7,800 (7,800)	0 9,300 (9,300)	0 3,900 (3,900)	0 7,500 (7,500)	0
1999	76,260 6,600 (82,860)	59,950 7,800 (67,750)	5,006 7,800 (12,806)	0 9,300 (9,300)	1,800 3,900 (5,700)	0 7,500 (7,500)	0
2000	76,260 6,600 (82,860)	59,950 7,800 (67,750)	5,006 7,800 (12,806)	0 9,300 (9,300)	1,800 3,900 (5,700)	0 7,500 (7,500)	0
2001	76,260 6,600 (82,860)	59,950 7,800 (67,750)	7,506 7,800 (15,306)	0 9,300 (9,300)	3,300 3,900 (7,200)	0 7,500 (7,500)	0
2002	76,260 6,600 (82,860)	59,950 7,800 (67,750)	7,506 7,800 (15,306)	0 9,300 (9,300)	3,300 3,900 (7,200)	0 7,500 (7,500)	0
2003	76,260 6,600 (82,860)	59,950 7,800 (67,750)	9,266 7,800 (17,066)	380 9,300 (9,680)	3,300 3,900 (7,200)	0 7,500 (7,500)	0
2004	78,607 6,600 (85,207)	64,476 7,800 (72,276)	28,338 7,800 (36,138)	2,100 9,300 (11,400)	3,300 3,900 (7,200)	0 7,500 (7,500)	0
2005	90,908 6,600 (97,508)	73,970 7,800 (81,700)	33,085 7,800 (40,882)	2,100 9,300 (11,400)	3,300 3,900 (7,200)	0 7,500 (7,500)	0

Note: First figure shows excise tax, the second figure in each cell starting in the year 1997 shows the CO_2 tax and the figures in brackets show total tax levied on energy products; coal—hard coal.

Table A.15. Overview of tax development in Sweden

	Petrol unleaded SEK/ 1,000l	Diesel (transport) SEK/1,000l	Light fuel oil SEK/ ton	Heavy fuel oil SEK/ ton	Coal SEK/ ton	Natural gas SEK/ 1,000 m³	Household Electricity SEK/MWh	Industry Electricity SEK/MWh
1988	2,330	660	778	778	305	308	72	50
1989	2,580	860	978	978	450	308	72	50
1990	2,920	960	1,078	1,078	460	350	72	50
1991	2,980	910	1,260	1,260	850	710	72	50
1992	2,950	810	1,260	1,260	850	710	72	50
1993	3,880	1,010	1,460	1,460	1,030	855	85	0
1994	3,910	2,260	1,519	1,519	1,071	889	88	0
1995	4,010	2,424	1,559	1,559	1,099	912	90	0
1996	4,160	2,530	1,644	1,644	1.167	979	97	0
1997	4,270	2,574	1,704	1,704	1,191	997	113	0
1998	4,470	2,672	1,801	1,801	1,236	1,033	152	0
1999	4,430	2,649	1,785	1,785	1,225	1,024	151	0
2000	4,470	2,922	1,801	1,801	1,236	1,033	162	0
2001	4,500	3,039	2,215	2,215	1,622	1,367	181	0
2002	4,620	3,121	2,505	2,505	1,865	1,575	198	0
2003	4,710	3,178	2,894	2,894	2,199	1,861	227	0
2004	4,790	3,331	3,330	3,330	2.572	2,183	241	5
2005	4,960	3,645	3,344	3,344	2,583	2,192	254	5
2006	4,990	3,665	3,362	3,539	2,597	2,204	261	5
2007	5,060	3,720	3,413	3,592	2,636	2,237	265	5

Note: NO_x charge and SO_2 tax are not considered.

Table A.16. Nominal vs. effective tax rates for the manufacturing industry

	Nominal tax rate				Effective tax rate—manufacturing industry			
	Light fuel oil SEK/ 1,000l	Heavy fuel oil SEK/ ton	Coal SEK/ ton	Natural gas SEK/ 1,000 m³	Light fuel oil SEK/ 1,000l	Heavy fuel oil SEK/ ton	Coal SEK/ ton	Natural gas SEK/ 1,000 m³
1990	1,078	1,078	460	350	1.078	1.078	460	350
1991	1,260	1,260	850	710	1,260	1,260	850	710
1992	1,260	1,260	850	710	1,260	1,260	850	710
1993	1,460	1,460	1.030	855	230	230	200	170
1994	1,519	1,519	1.071	889	239	239	208	177
1995	1,559	1,559	1.099	912	246	246	214	181
1996	1,644	1,644	1,167	979	264	264	229	197
1997	1,704	1,704	1,191	997	263	263	229	197
1998	1,801	1,801	1,236	1,033	529	529	460	396
1999	1,785	1,785	1,225	1,024	525	525	456	393
2000	1,801	1,801	1,236	1,033	529	529	460	396
2001	2,215	2,215	1.622	1,367	535	535	466	401
2002	2,505	2,505	1,865	1,575	539	539	469	404
2003	2,894	2,894	2,199	1,861	544	544	473	407
2004	3,330	3,330	2,572	2,183	546	546	475	409
2005	3,344	3,344	2,583	2,192	548	548	477	410
2006	3,362	3,539	2,597	2,204	551	580	479	413
2007	3,413	3,592	2.636	2,237	605	605	527	453

Table A.17. The UK energy tax rates on mineral oil products

	Petrol unleaded UK£/litre	Diesel (transport) UK£/litre	Gas oil UK£/ litre	Fuel oil UK£/ litre	Heavy fuel oil UK£/ litre	Electricity— fossil fuel levy *ad valorem tax (in %)*
1988	0.1842	0.1729	0.011	0.0077	0.00778	
1989	0.1772	0.1729	0.011	0.0077	0.00778	
1990	0.192	0.207	0.012	0.008	0.00823	8.8
1991	0.219	0.219	0.013	0.009	0.00897	9.1
1992	0.233	0.227	0.013	0.009	0.00953	9.1
1993	0.256	0.248	0.015	0.010	0.0105	8.3
1994	0.285	0.279	0.017	0.012	0.016	8.3
1995	0.316	0.316	0.022	0.017	0.0166	8.3
1996	0.345	0.345	0.023	0.018	0.0181	3.1
1997	0.385	0.386	0.025	0.020	0.0194	1.9
1998	0.431	0.438	0.028	0.021	0.02	0.8
1999	0.467	0.493	0.030	0.026	0.0218	0.2
2000	0.486	0.516	0.031	0.027	0.0265	0.3
2001	0.480	0.518	0.031	0.027	0.0274	0.4
2002	0.488	0.518	0.031	0.027	0.0274	0.0
2003	0.492	0.522	0.039	0.036	0.0274	0.0
2004	0.502	0.533	0.043	0.039	0.0382	0.0
2005	0.506	0.537	0.056	0.052	0.0482	0.0
2006	0.502	0.533	0.0644	0.0644	0.0064	0.0
2007	0.515	0.547	0.0769	0.0969	0.073	0.0

Table A.18. The Climate Change Levy (only levied on energy consumption by business)

	Only business use			
	Natural gas UK£/kWh	Electricity UK£/kWh	Coal UK£/kWh	lpg UK£/kWh
2001	0.0015	0.0043	0.0015	0.0007
2002	0.0015	0.0043	0.0015	0.0007
2003	0.0015	0.0043	0.0015	0.0007
2004	0.0015	0.0043	0.0015	0.0007
2005	0.0015	0.0043	0.0015	0.0007
2006	0.0015	0.0043	0.0015	0.0007
2007	0.00154	0.0041	0.0154	0.000718

Table A.19. Overview of tax rates on light fuel oil—nominal versus effective rates (industry): in EUR/1,000 litres

	Nominal tax rates							Effective tax rates								
	Dk	Fin	Ger	NL	Slov	Sw	UK	Space heating Dk	Light process-with agreement Dk	Heavy process-with agreement Dk	Fin	Ger	NL	NL	Sw	UK
1988	221.4		8.2	52.4		107.5	11.7					8.2	52.4	52.4	107	11.7
1989	218.6		28.1	50.8		137.7	11.5					28.1	50.8	50.8	138	11.5
1990	223.9	3.4	28.4	52.9	20%	143.4	11.5				3.4	28.4	52.9	52.9	143.4	11.5
1991	222.5	3.5	39.0	53.6	20%	168.4	12.7				3.5	39.0	53.6	53.6	168.4	12.7
1992	225.4	3.5	39.6	62.0	20%	167.3	12.7				3.5	39.6	62.0	62.0	167.3	12.7
1993	231.9	14.1	41.2	65.0	20%	160.1	13.3				14.1	41.2	65.0	65.0	25.2	13.3
1994	233.4	20.5	41.7	66.1	20%	165.8	15.4				20.5	41.7	66.1	66.1	26.1	15.4
1995	240.2	30.2	42.7	62.0	20%	167.1	20.2				30.2	42.7	62.0	62.0	26.3	20.2
1996	239.2	30.2	41.9	74.0	20%	193.1	22.4		18.3	1.1	30.2	41.9	74.0	60.8	31.0	22.4
1997	235.2	48.8	40.7	71.6	14.4	197.0	28.5		18.0	1.1	48.8	40.7	71.6	59.2	30.4	28.5
1998	262.7	52.0	40.9	96.8	41.9	202.0	31.6	262.7	18.0	1.1	52.0	40.9	96.8	58.7	59.3	31.6
1999	264.9	63.7	61.4	116.8	66.1	202.7	39.0	264.9	21.1	1.1	63.7	45.0	116.8	59.3	59.6	39.0
2000	268.3	63.7	61.4	138.6	62.5	213.3	44.6	268.3	24.6	1.1	63.7	45.0	138.6	59.5	62.6	44.6
2001	275.1	63.7	61.4	186.2	70.5	239.3	44.1	275.1	24.6	1.1	63.7	45.0	186.2	59.7	57.8	44.1
2002	282.6	63.7	61.4	196.9	67.7	273.4	43.6	282.6	24.7	1.1	63.7	45.0	196.9	60.2	58.9	43.6
2003	282.6	67.1	61.4	197.0	73.0	317.2	51.3	282.6	24.7	1.1	67.1	53.2	197.0	60.7	59.6	51.3
2004	282.3	67.1	61.4	205.2	153.5	365.0	57.5	282.3	24.7	1.1	67.1	53.2	205.2	61.0	59.8	57.5
2005	282.3	67.1	61.4	206.3	170.3	366.5	77.0	282.3	24.7	1.1	67.1	53.2	206.3	61.6	59.8	77.0
2006	281.5	67.1	61.4	207.7		356.2	94.3	281.5	24.6	1.1	67.1	53.2	207.7		59.5	94.3
2007	278.3	70.6	61.4	208.2		369.0	141.7	278.3	24.6	1.1	70.6	45.0	207.7		60.5	141.7

Note: Effective tax rates: special tax provisions are considered where applicable! NL—second row is valid for consumption of light fuel exceeding the taxable event (ceiling is 159,000 litres per annum).

[3] The following tables compare the development of nominal tax rates and the taxes levied on industry taking into account special tax provisions, such as tax exemptions. Special refund schemes, such as the ones implemented for example in Finland and Sweden, are not considered. These tables can only be seen as indicative as the reality may look different.

Table A.20. Overview of tax rates on heavy fuel oil—nominal versus effective rates (industry) in EUR/ton

| | Nominal tax rates | | | | | | | Effective tax rates-industry | | | | | | | |
	Dk	Fin	Ger	NL	Slov	Sw	UK	Space heating Dk	Light process-with agreement Dk	Heavy process-with agreement Dk	Fin	Ger	NL	Sw	UK
1988	249.1	0	7.2	16.8		107.5	11.9				0	7.2	16.8	107	11.9
1989	246.0	0	14.5	17.1		137.7	11.7				0	14.5	17.1	138	11.7
1990	251.9	3.4	14.6	20.3	32%	143.4	11.7				3.4	14.6	20.3	143.4	11.7
1991	250.3	3.5	14.6	22.5	32%	168.4	13.0				3.5	14.6	22.5	168.4	13.0
1992	253.5	3.5	14.9	29.2	32%	167.3	13.0				3.5	14.9	29.2	167.3	13.0
1993	260.9	11.2	15.5	29.9	32%	160.1	13.6				11.2	15.5	29.9	25.2	13.6
1994	262.6	19.8	15.6	30.2	32%	165.8	20.7				19.8	15.6	30.2	26.1	20.7
1995	270.2	31.2	16.0	31.7	32%	167.1	20.2				31.2	16.0	31.7	26.3	20.2
1996	269.0	31.2	15.7	31.1	32%	193.1	22.5		21.7	1.3	31.2	15.7	31.1	31.0	22.5
1997	264.6	37.2	15.3	30.1	17.2	197.0	28.3		21.4	1.3	37.2	15.3	30.1	30.4	28.3
1998	264.0	44.0	15.2	30.0	49.9	202.0	29.9	264.0	21.3	1.3	44.0	15.2	30.0	59.3	29.9
1999	299.9	54.0	15.3	30.5	48.0	202.7	33.5	299.9	25.0	1.3	54.0	15.2	30.5	59.6	33.5
2000	304.5	54.0	15.3	30.8	45.4	213.3	44.0	304.5	29.2	1.3	54.0	15.8	30.8	62.6	44.0
2001	311.3	54.0	17.9	31.0	42.8	239.3	44.5	311.3	29.2	1.3	54.0	15.8	31.0	57.8	44.5
2002	320.3	54.0	17.9	31.5	41.1	273.4	44.1	320.3	29.3	1.3	54.0	15.8	31.5	58.9	44.1
2003	320.3	56.8	25.0	32.1	41.4	317.2	40.0	320.3	29.3	1.3	56.8	19.5	32.1	59.6	40.0
2004	319.9	56.8	25.0	32.5	48.4	365.0	56.9	319.9	29.3	1.3	56.8	19.5	32.5	59.8	56.9
2005	319.9	56.8	25.0	32.5	47.5	366.5	71.8	319.9	29.2	1.3	56.8	19.5	32.5	59.8	71.8
2006	319.0	56.8	25.0	32.5		374.9	94.7	319.0	29.2	1.3	56.8	19.5	32.5	59.5	94.9
2007	319.5	59.6	25.0	32.5		388.3	107.9	319.5	29.2	1.3	59.6	15.0	32.5	60.4	107.9

Note: Effective tax rates: special tax provisions are considered where applicable!

Table A.21. Overview of tax rates on coal—nominal versus effective rates (industry) in EUR/ton

	Nominal tax rates							Effective tax rates-industry								
	Dk	Fin	Ger	NL	Slov	Sw	UK-CCL	Space heating Dk	Light process-with agreement Dk	Heavy process-with agreement Dk	Fin	Ger	NL	Slov	Sw	UK-CCL 80%
1988	84.9		0	2.5		42.1	0				0	0	2.5		42	0
1989	95.0		0	2.5		63.4	0				0	0	2.5		63	0
1990	97.3	2.7	0	5.5	5%	61.2	0				2.7	0	5.5		61.2	0
1991	96.7	2.8	0	9.1	5%	113.6	0				2.8	0	9.1		113.6	0
1992	119.3	2.8	0	10.9	5%	112.9	0				2.8	0	10.9		112.9	0
1993	122.8	5.6	0	10.4	5%	112.9	0				5.6	0	10.4		21.9	0
1994	123.6	11.3	0	10.5	5%	116.9	0				11.3	0	10.5		22.7	0
1995	138.1	19.5	0	11.1	5%	117.8	0				19.5	0	11.1		22.9	0
1996	149.7	19.5	0	10.9	5%	137.1	0				19.5	0	10.9		26.9	0
1997	159.3	28.4	0	10.6	13.9	137.7	0		16.4	1.0	28.4	0	10.6		26.4	0
1998	170.9	33.4	0	10.5	40.3	138.6	0	170.9	16.2	1.0	33.4	0	10.5		51.6	0
1999	200.7	41.4	0	10.8	38.7	139.1	0	200.7	16.1	1.0	41.4	0	10.8		51.8	0
2000	206.9	41.4	0	11.0	36.6	146.4	0	206.9	18.9	1.0	41.4	0	11.0		54.5	0
2001	213.6	41.4	0	11.2	34.5	175.3	19.5	213.6	22.1	1.0	41.4	0	11.2		50.3	3.9
2002	224.3	41.4	0	12.0	33.2	203.6	19.3	224.3	22.1	1.0	41.4	0	12.0		51.2	3.9
2003	224.3	43.5	0	12.0	32.1	241.0	17.5	224.3	22.1	1.0	43.5	0	12.0		51.8	3.5
2004	224.1	43.5	0	12.3	31.9	281.9	17.9	224.1	22.1	1.0	43.5	0	12.3		52.0	3.6
2005	224.1	43.5	0	12.5	31.3	283.1	17.9	224.1	22.1	1.0	43.5	0	12.5		52.1	3.6
2006	223.5	43.5	0	12.5		275.1	17.8	223.5	22.1	1.0	43.5	0	12.5		50.8	3.6
2007	223.7	44.7	8.4	12.8		285.0	18.2	223.7	22.1	1.0	44.7	8.4	12.8		52.6	3.6

Note: Effective tax rates: special tax provisions are considered where applicable! UK—CCL: 80 per cent reduction for energy-intensive industries.

Table A.22. Overview of tax rates on natural gas—nominal versus effective rates (industry) in EUR/1,000 m³

| | Nominal tax rates | | | | | | | | Effective tax rates-industry | | | | | | |
| | Consumption 5,001–170,000 | | | | | | | Space heating | Light process with agreement | Heavy process with agreement | Consumption 1M–10M | | | | |
	Dk	Fin	Ger	NL	Slov	Sw	UK-CCL	Dk	Dk	Dk	Fin	Ger	NL	Sw	UK-CCL (80%)
1988	0	0	0	0		42.5	0				0.0	0	0	42.5	0
1989	0	2.0	13.6	0		43.4	0				2.0	13.6	0	43.4	0
1990	0	2.0	13.7	0	5%	46.5	0				2.0	13.7	0	46.5	0
1991	0	2.0	19.0	0	5%	94.9	0				2.0	19.0	0	94.9	0
1992	0	2.0	19.3	0	5%	94.3	0				2.0	19.3	0	94.3	0
1993	0	4.0	20.1	0	5%	93.8	0				4.0	20.1	0	18.6	0
1994	0	11.0	20.3	0	5%	97.1	0				11.0	20.3	0	19.3	0
1995	0	9.0	20.8	10.3	5%	97.7	0				9.0	20.8	10.3	19.4	0
1996	31	9.0	20.4	25.0	5%	115.0	0		14.9	0.9	9.0	20.4	10.1	23.2	0
1997	194	12.0	19.9	38.7	7.2	115.2	0	225.4	14.7	0.9	12.0	19.9	9.7	22.7	0
1998	225	14.0	19.8	52.6	20.9	115.9	0	227.3	14.7	0.9	14.0	19.8	9.7	44.4	0
1999	227	17.0	37.7	82.5	29.4	116.3	0	244.2	17.2	0.9	17.0	23.5	10.0	44.6	0
2000	244	17.0	37.7	62.1	27.8	122.3	0	292.5	20.1	0.9	17.0	23.5	10.2	46.9	0
2001	293	17.0	37.7	66.5	33.2	147.7	26.1	301.5	20.1	0.9	17.0	23.5	10.3	43.3	5.2
2002	301	17.0	37.7	68.9	31.8	171.9	25.8	301.5	20.1	0.9	17.0	23.5	11.0	44.1	5.2
2003	301	18.0	59.6	71.0	30.8	204.0	23.5	301.1	20.1	0.9	18.0	43.9	11.0	44.6	4.7
2004	301	18.0	59.6	72.7	30.6	239.3	23.9	301.1	20.1	0.9	18.0	43.9	11.3	44.8	4.8
2005	301	18.0	59.6	101.9	30.0	240.2	23.9	301.1	20.1	0.9	18.0	43.9	11.5	44.8	4.8
2006	300.3	18.0	59.6	123.8		233.5	23.8	300.3	20.1	0.9	18.0	43.9	11.6	43.7	4.8
2007	300.7	19.0	59.6	134.2		241.8	24.4	300.7	20.1	0.9	19.0	35.8	11.8	45.3	4.9

Note: effective tax rates: special tax provisions are considered where applicable.! UK—CCL 80 per cent reduction for energy-intensive industries

Table A.23. Overview of tax rates on electricity—nominal versus effective rates (industry) in EUR/MWh

| | Nominal tax rates Consumption: 800–10,000 | | | | | | | Effective tax rates–industry Consumption: 50,000–10 M | | | | | |
	Denmark EUR/MWh	Finland EUR/MWh	Germany EUR/MWh	Netherlands EUR/MWh	Slovenia EUR/MWh	Sweden EUR/MWh	UK-CCL EUR/MWh	Denmark EUR/MWh	Finland EUR/MWh	Germany EUR/MWh	Netherlands EUR/MWh	Sweden EUR/MWh	UK-CCL (80%) EUR/MWh
1988	41.0	0	7.4			9.9	0		0	4.8		6.9	0
1989	40.8	0	8.8			10.1	0		0	5.6		7.0	0
1990	42.0	0	8.6		5%	9.6	0		0	5.5		6.6	0
1991	41.7	0	8.6		5%	9.6	0		0	5.4		6.7	0
1992	47.4	0	8.4		5%	9.6	0		0	5.2		6.6	0
1993	48.7	0	8.8		5%	9.3	0	6.6	0	5.3		0	0
1994	53.1	0	10.3		5%	9.6	0	6.6	0	5.9		0	0
1995	58.7	0	10.6		5%	9.6	0	6.8	0	6.0		0	0
1996	62.5	0	0	13.8	5%	11.4	0	6.8	0	0	0	0	0
1997	66.8	5.6	0	13.3	0	13.1	0	8.0	2.4	0	0	0	0
1998	75.5	5.6	0	13.3	0	17.0	0	9.3	3.4	0	0	0	0
1999	78.1	6.9	10.2	22.5	0	17.1	0	10.8	4.2	2.0	1.0	0	0
2000	85.3	6.9	12.8	37.2	0	19.2	0	12.1	4.2	2.6	2.2	0	0
2001	87.4	6.9	15.3	58.3	0	19.6	6.9	12.1	4.2	3.1	5.9	0	1.4
2002	89.6	6.9	17.9	60.1	0	21.6	6.8	12.1	4.2	3.6	6.1	0	1.4
2003	89.6	7.2	20.5	63.9	0	24.9	6.2	12.1	4.4	12.3	6.3	0	1.2
2004	89.5	7.2	20.5	65.4	0	26.4	6.3	12.1	4.4	12.3	6.5	0.5	1.3
2005	89.5	7.2	20.5	69.9	0	27.8	6.3	12.1	4.4	12.3	8.6	0.5	1.3
2006	89.3	7.3	20.5	70.5		28.2	6.3	12.2	4.4	12.3	9.4	0.5	1.3
2007	89.4	7.43	20.5	71.6		28.6	6.4	12.2	2.3	12.3	10.2	0.5	1.3

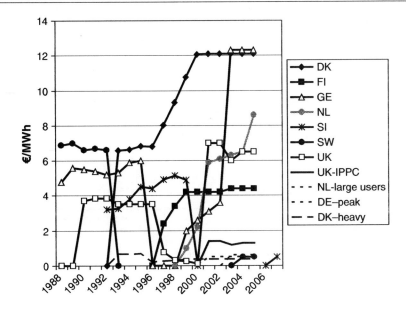

Figure A.5. Electricity tax rate for industrial end-users

Index

Index